SpringerBriefsinM olecularSc ience

GreenC hemistryforSus tainability

SeriesEditor

Sanjay K. Sharma

Forfurthe rv olumes:
http://www.springer.com/series/10045

TingyueG u

Editor

Green Biomass Pretreatment for Biofuels Production

 Springer

Editor
Tingyue Gu
Department of Chemical and Biomolecular Engineering
Ohio University
Athens, OH
USA

ISSN2212-9898
ISBN978-94-007-6051-6   ISBN978- 94-007-6052-3   (eBook)
DOI10.1007/978-94-007-6052-3
Springer Dordrecht Heidelberg New York London

LibraryofC ongressC ontrolN umber:2012955749

Springer is part of Springer Science+Business Media (www.springer.com)

Preface

In order to counter the increasing global demand for energy, a multifaceted approach is needed. Renewable bioenergy produced from various types of biomass, especially lignocellulosic biomass, is an important source of energy supply. Such biomass sources can be agricultural wastes or direct harvests from high-yield energy corps. Lignocellulosics are recalcitrant by nature's design. Fermentable sugars must be liberated from the biomass before ethanol fermentation. Pretreatment is necessary and it is often an expensive part of the overall lignocellulosic ethanol process, which is critical to the economic feasibility of lignocellulosic ethanol. A good pretreatment method makes the subsequent enzyme hydrolysis step much more effective while minimizing the formation of fermentation inhibitors. This book emphasizes on potential on-farm and tactical mobile applications that use green pretreatment and processing methods without the need for on-site waste treatment. The first chapter in this book, discusses plant cell wall structures and their impact on pretreatment. Each subsequent chapter is dedicated to one pretreatment methods. These methods include mechanical pretreatment, biological pretreatment, hydrothermal pretreatment (including steam explosion), supercritical CO_2 explosion pretreatment, and ionic liquid pretreatment. Because there a huge variety of different lignocellulosic biomass types that are potential targets, no single pretreatment is expected to be the universal choice. Some of the pretreatment methods are niche applications that are particularly suitable for the desired small-scale on-farm or mobile operations that eliminate the need to transport bulky raw biomass feedstock. Mobile or tactical bioenergy production is already a feasible option for military forward operating bases in a war zone where fuel costs can be ten times higher than normal.

Athens,O H,U SA

Tingyue Gu

Contents

Contributors

Grzegorz Brudecki Department of Agricultural and Biosystems Engineering, SouthD akotaSt ateU niversity,B rookings,SD 57007 ,U SA

Meiqiang Cai College of Environmental Science and Engineering, Zhejiang GongshangU niversity,H angzhou310035 ,C hina

Iwona Cybulska Chemical Engineering Program, Masdar Institute of Science and Technology, AbuD habi,U nited ArabEmira tes

Ahmed Faik Department of Environmental and Plant Biology, Ohio University, Athens,O H45701 ,U SA

Tingyue Gu Department of Chemical and Biomolecular Engineering, Ohio University, Athens,O H45701 ,U SA

Chinnadurai Karunanithy Food and Nutrition, University of Wisconsin-Stout, Menomonie, WI54751 ,U SA

Hanwu Lei Department of Biological Systems Engineering, Washington State University,R ichland, WA99354 ,U SA

Yebo Li Department of Food, Agricultural, and Biological Engineering, Ohio State University/OARDC, Wooster,O H44691 ,U SA

Jian Luo National Key Lab of Biochemical Engineering, Institute of Process Engineering, 100190B eijing,C hina

Kasiviswanathan Muthukumarappan Department of Agricultural and Biosystems Engineering,SouthD akotaSta teU niversity,B rookings,SD 57007 ,U SA

Caixia Wan Department of Environmental Science and Engineering, Fudan University,200433 Sha nghai,C hina

Chapter 1
"PlantC ell WallStr ucture-Pretreatment" the Critical Relationship in Biomass Conversion to Fermentable Sugars

AhmedF aik

Abstract One of the targeted research areas in implementing alternative renewable energy production is the improvement of the yield and quality of plant biomass, which consists mostly of plant cell walls (called lignocellulosic biomass). Sugars from plant biomass can be used to produce bioethanol through fermentation, but can also be used to make other hydrocarbons via direct pyrolysis or gasification. However, the conversion of lignocellulosic biomass to fermentable sugars is far from optimal due to the lack of efficient pretreatment processes, simply because the exact composition and the manner in which different cell wall components interact between each other strongly influence energy recovery. Pretreatment techniques can be grouped into three distinct categories: *physical* (mechanical), *biological*, and *chemical pretreatments*. Currently, pretreatment step is the most costly step in the whole process of biofuel production, and there is a positive correlation between cell walls recalcitrance and the costs in biofuel production. Because there are many different kinds of plant biomass, no single pretreatment method is expected to be the preferred universal choice. Furthermore, there is recurring debate about "food or fuel" balance, and the emerging picture from this debate is that there is a need for domestication of several feedstock crops, because none of the current available feedstock crops have all the requirements to balance our food and fuel needs. The ultimate goal is to elucidate the key structural elements of lignocellulosic biomass that would allow a balance between biofuel production, carbon sequestration, and land management. This chapter will describe the composition and structural aspects of plant cell walls found in plant biomasses, and their impact on pretreatment of biomass.

Keywords Lignocellulosic biomass • Cell wall • Pretreatment • Biofuels • Energy balance • Hemicellulose • Cellulose • Xylan • Bioethanol • Lignin • *Miscanthus* • Switchgrass • Oil • Fermentation • Saccharification • Sustainability • Green energy • Basidiomycetes • *Saccharomyces cerevisiae* • Yeast

A. Faik(✉)
DepartmentofEn vironmentala ndPl antB iology,O hioU niversity, Athens,O H45701 ,U SA
e-mail:f aik@ohio.edu

T. Gu(e d.), *Green Biomass Pretreatment for Biofuels Production*,
SpringerBriefs in Green Chemistry for Sustainability,
DOI:10.1007/978-94- 007-6052-3_1,© The Author(s)2013

1.1 Introduction

Until recently (2002), fossil fuels supplied ~86 % of the energy consumed in the United States, and more than half (62 %) of that was imported. To achieve its energy needs and independence from foreign oil, the United States set a goal to substitute 20 % of its gasoline usage with alternative fuels by 2022, which would correspond to an increase in the annual alternative fuel production of up to 35 billion gallons. On the other hand, it is important to understand that no matter how many new oil fields are (or will be) discovered, oil is a finite and nonrenewable resource (Campbell 2006). Therefore, alternative renewable energy production needs to be implemented. One of the targeted research areas to reach this goal is the improvement of the yield and quality of plant biomass, which consists mostly of plant cell walls (called lignocellulosic biomass). These cell walls are the result of capturing and converting solar energy into energy-rich polymers (carbohydrates) via carbon fixation during photosynthesis. Although the annual solar energy reaching Earth's surface is evaluated to be ~5.5 × 10^{24} J, which is 12,000 times the annual global energy demand (Larkum et al. 2011), the amount of carbon annually fixed is estimated to 10^{11}–10^{12} tons, which if converted to biofuels would correspond to only 10 times our energy need (Hall 1979). Therefore, biofuels production is not limited by energy input, but rather by other factors such as efficient use of land and nutrient to efficiently harvest the light and fix the carbon into biomass, and the conversion of this biomass to sugars that are then fermented to biofuels. Another alternative for efficient light harvesting and usage as renewable energy could be the artificial photosynthetic systems (e.g. photochemical panels); however, these systems are currently limited to the production of only hydrogen (from phytolysis: hydrolysis of water to hydrogen and oxygen) (Melis et al. 2000; Hemschemeier et al. 2009). Biomass sugars can be used to make bioethanol (through pretreatment and fermentation), but can also be used to make other hydrocarbons via direct pyrolysis or gasification (Regalbuto 2009).

The conversion of lignocellulosic biomass to fermentable sugars is far from optimal due to the lack of efficient pretreatment processes, simply because the exact composition and the manner in which different cell wall components interact between them strongly influence energy recovery (as fermentable sugars). Thus, biomass pretreatment is a critical step in this process. Figure 1.1 summarizes the average values of cellulose, hemicellulose, and lignin content (as % dry weight) in most promising lignocellulosic biomass such as wheat straw (Lynd et al. 1999; Kabel 2007; Carvalheiro et al. 2009; Thomsen et al. 2008), barley straw (Saha and Cotta 2010; Persson et al. 2009), corn stover (Kaar and Holtzapple 2000; Kim et al. 2003; Zhang et al. 2007; Chundawat et al. 2007), hardwood (Kim et al. 2000; Howard 2003; Sjostrom 1993), softwood (Howard 2003; Miller and Hester 2007; Mabee et al. 2006), *Miscanthus* (Majid et al. 2004), switchgrass (Lynd et al. 1999), and rice straw (Jin and Chen 2007).

Currently, U.S. generate ~1.3 billion tons of plant biomass, which should allow the production of ~130 billion gallons of bioethanol and other biofuel molecules under optimal conditions (Somerville 2006). Bioethanol obtained from cereal grains and biodiesel obtained from soybean and rapeseed are considered the first-generation biofuels, while biofuels from nonfood annual crops (e.g., sorghum,

Fig.1.1 Comparison of cellulose. hemicellulose, and lignin contents in eight plant lignocellulosic biomasses: Switchgrass, *Miscanthus*, softwood, hardwood, corn stover, and straw from rice, barley, and wheat (see text for references)

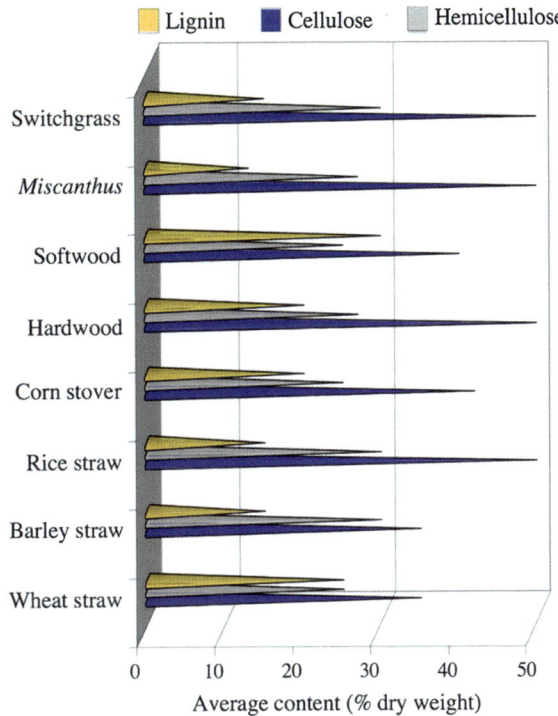

switchgrass, and *Miscanthus*, Fig. 1.1) are considered the second generation, also called lignocellulosic biofuels (Yuan et al. 2008). Products with energy content closer or higher to petroleum and diesel (e.g., the high-chain alcohols such as alkanes and butanol) are considered as the next-generation biofuels (Duree 2007). However, in the U.S., even if all biomass currently available were converted to biofuels, the U.S. energy needs would still not be met (Hill et al. 2006).

This chapter will describe the structural aspects of plant cell walls found in plant biomasses, and their impact on biofuel yields. The ultimate goal is to elucidate the key structural elements of lignocellulosic biomass that would allow a balance between biofuels production, carbon sequestration, and land management. With scientific progress in genomics and genetic, it would be possible to engineer or domesticate some plant species that would help achieve this goal.

1.2 HistoryandC ontext

The idea of using biofuels is not new (Muller 1974). It started in 1893 when Rudolf Diesel, a thermal engineer, invented the pressure-ignited heat engine that ran on its own power. In 1900, Diesel used peanut oil to demonstrate that his engine can run on vegetable oil ("Paris Exhibition of Alcohol Consuming

Devices," <u>Scientific American</u>, Nov. 16, 1901). At that time, Diesel said: *The use of vegetable oils for engine fuels may seem insignificant today, but such oils may become, in the course of time, as important as petroleum...* At the same time (1908) in the U.S., Henry Ford designed his Model T to run on ethanol produced from corn. However, prohibition of use of ethanol in all forms in 1919 in U.S. brought this adventure to a halt, until 1970s, when the Clean Air Act was adopted by the environmental protection agency (EPA) to promote the use of biofuels. By 1944, three-quarters of the tires and other rubber products were produced using ethanol. In the current context, there is an intense discussion about the relative merits of a variety of alternative biomass sources for biofuels. This discussion is driven by three main questions: how much of the biomass would be needed if the country were to switch to biofuels? does it take more energy to produce a biofuel than it actually contains? And finally, would food production be sacrificed to fuel needs? The purpose of these questions is to raise awareness about the efficient use of the available land, and the balance between generating renewable biomass and providing food to an increasing world population (projected to reach up to 10.5 billion by 2050).

1.3 Plant Cell Wall Composition and Current Models: When the Fine Structural Variation Matters

Biofuel yields depend on the variations in the structural and chemical composition of plant cell walls. These structurally complex cell walls allowed the plants to succeed in colonizing land, and stand tall for better capturing light energy. Cell walls were designed to be strong and serve for a tall growth and protection/defense under various land climates. Any abnormality in these cell walls produces defect and/or halt of the growth of the plants (Reiter et al. 1993). It is this strength in cell wall structure that makes it recalcitrant to conversion to biofuel.

1.3.1 General Chemical Composition of Plant Cell Walls

All plant cell walls are made of carbohydrate polymers (microfibrils of cellulose, hemicellulose, and pectin) and other noncarbohydrate polymers such as proteins, and lignin (produced from phenylpropanoid units). Carbohydrate polymers and lignin are insoluble polymers and must be synthesized from soluble subunits (within plant cells) before assembly and deposition (outside of the cells) as insoluble but functional plant cell walls. The chemical composition of the cell walls has been extensively studies since the early 1970s. Figure 1.2 summarizes the general chemical structure and composition of cellulose, hemicellulose, and pectin along with their building units.

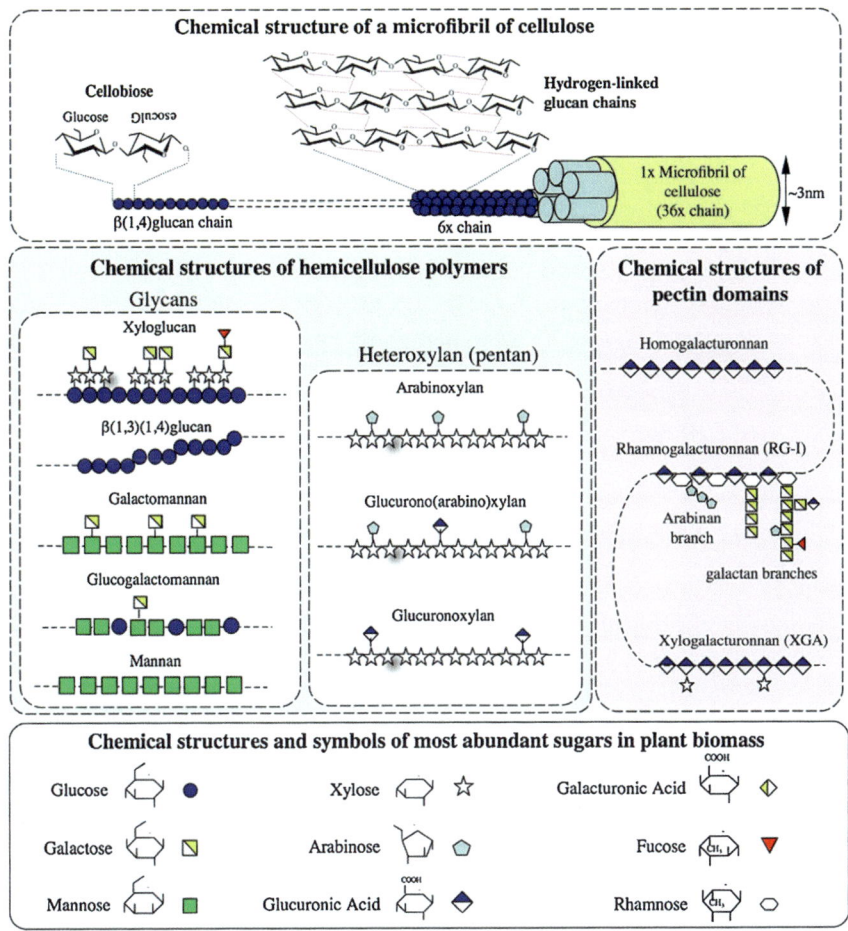

Fig. 1.2 Chemical structure of the building blocks of plant cell wall polymers. *Bottom panel* shows the presentation of the structures and symbols of these building blocks (monomers). *Upper panels* show the representation of the polymers, namely cellulose, hemicellulose, and pectin

1.3.1.1 Cellulose

Cellulose is made of repeating β(1,4)-linked cellobiose units. Cellobiose is a disaccharide formed of two D-glucose residues themselves linked via a β(1,4)-linkage but one residue is right side up and the other is upside down (Fig. 1.2). Each β(1,4)-D-glucan chain is an assembly of up to 10,000 glucose residues depending of the species and tissues (Klemm et al. 2005). About 36 of these β(1,4)-D-glucan chains are orderly assembled and held together via interchain hydrogen bonds to form a long rod-like microfibrils (Fig. 1.2). The assembly of these cellulose microfibrils occurs concomitantly with the synthesis of individual glucan chains by a symmetrical multiprotein enzyme complex called rosette (Mueller and Brown 1980;

McCann and Roberts 1991; Carpita and Gibeaut 1993; Brown et al. 1996; Ding and Himmel 2006; Wightman and Turner 2010). The diameter of each microfibril ranges from 3 to 68 nm depending of the biomass source (Brown et al. 1996; Jarvis 2003; Tsekos 1999), and up to 5 μm in length. Because of the crystalline structure of cellulose microfibrils (due to tight hydrogen-bounding between the chains), dissociation of the glucan chains (to become amorphous) requires harsh conditions such as heating at 320 °C under a pressure of 25 MPa (Deguchi et al. 2006). Because cellulose represents up to 50 % of the weight of plant lignocellulosic biomass (Fig. 1.1), and is constituted of glucose (Glc) that is easily fermented, all pretreatment methods were optimized to maximize the release of Glc residues and minimize its conversion to other byproduct before fermentation.

1.3.1.2 Hemicellulose

Hemicellulose is a generic name for a heterogeneous group of plant cell wall polymers that have typically β(1,4)-D-glycan backbones (e.g., xylan, mannan, xyloglucan) branched with one monosaccharide and/or small oligosaccharides (Fig. 1.2). They share another characteristic that consists in their solubility in low concentration alkali solutions (e.g., 0.1–0.4 M KOH or NaOH). These hemicellulosic polymers do not form crystalline structure (compared to cellulose). Plant biomass for biofuels contains the following hemicellulose (with varying amounts): xyloglucan (XyG), β(1,3)(1,4)-D-gucan (called also mixed-linkage glucan, MLG), β(1,4)-D-xylan, and β(1,4)-D-mannan. Figure 1.2 summarizes the simplified diagrammatical presentation of the chemical structure of these polymers.

Xylan is the third most abundant polymer on earth after cellulose and chitin (not present in plant cell walls). It consists of various types of polymers that can be simple homopolymers such as β(1,4)- and β(1,3)-D-xylan; or heteropolymers such as glucuronoxylan (GX) found in the wood of dicots, glucurono(arabino)xylan (GAX) found in softwoods and vegetative tissues of grasses, and neutral arabinoxylan (AX) found in the endosperm of cereal grains (Ebringerova and Heinze 2000, Faik 2010). All these xylan types have a backbone of β(1,4)-linked D-xylose (Xyl) residues in common. This xylan backbone is substituted at the C-2 and/or C-3 positions, mainly with α-arabinofuranose (Ara*f*) residues, or to a lesser extent, with α-D-glucuronyl or *O*-methyl-α-D-glucuronyl uronic acid (GlcA) residues on the C-2 position, depending on the tissues and species (Fig. 1.2). The fact that Ara and GlcA substitutions can be esterified with ferulic, acetic, or *p*-coumaric acid groups, a xylan chain can be cross-linked to either lignin or to another xylan chain (Fry 1986; Watanabe and Koshijima 1988; Ebringerova and Heinze 2000; Saulnier et al. 1995). This cross-linking is the main contributor to the recalcitrance of xylan-rich plant biomasses (e.g., grasses and trees), and can have a great impact on biofuel production, especially if we consider that GX and GAX are the predominant (up to 50 % w/w) hemicellulose in the most promising plant biomass for biofuel production (e.g., grasses and trees).

The other types of hemicellulose are mostly hexose-based (6-carbon sugars) polymers, which include XyG, MLG, and mannan (Fig. 1.2). Although both XyG and mannan are considered as cell wall storage polysaccharides found in seeds of many taxonomically important plant groups, XyG is also the major hemicellulose (up to 30 % w/w) in cell walls of vegetative tissue of many herbaceous dicot plants and non-graminaceous monocots (Carpita and Gibeaut 1993; Buckeridge 2010). XyG has a regular structure with a backbone composed of β(1,4)-linked D-Glc residues, and up to 75 % of these Glc residues are substituted at C-6 positions with xylosyl residues, galactosyl-β(1,2)-D-xylosyl disaccharide, or L-fucosyl-α(1,2)-D-galactosyl-β(1,2)-D-xylosyl trisaccharide side chains (Fig. 1.2). Mannan and glucomannan can be substituted with single units of α(1,6)-linked D-galactosyl (Gal) to the main chain (Fig. 1.2).

In the case of MLG, the uniform β(1,4)-glucan backbone structure is interrupted at regular intervals (up to 5 Glc residues) by β(1,3)-linkages, which introduces kinks in the polymer and makes it more flexible and soluble in water (Miller and Fulcher 1995; Manthey et al. 1999; Burton and Fincher 2009). In contrast to xylan, both XyG and MLG are present in low amounts in mature plant tissues used for biofuels, but are major hemicelulose (up to 30 % w/w) in the primary cell walls (developing plants). Thus, they have less impact on biofuel production from plant biomass. However, because of their importance in plant development, they have direct effect on plant biomass yields.

1.3.1.3 Pectin

Pectin is a group of polysaccharides that are characterized by their richness in D-galacturonic acid (GalA) residues, a negatively charged monosaccharide that makes pectin easy to solubilize in hot water. It is believed that pectin polymers contain three structurally different domains consisting of homogalacturonan (HG) domain, rhamnogalacturonan II (RG-II) domain, and RG-I domain (Fig. 1.2). HG domain is a linear and unbranched chain of α(1,4)-linked GalA. When some of GalA residues of this HG domain are substituted with β(1,4)-linked D-Xyl residues on C-3 positions, the domain is called xylogalacturonan (XGA) (Zandleven et al. 2006). RG-II domain has a backbone similar to HG (at least eight GalA residues), but bearing on the C-3 positions of GalA residues complex oligosaccharide side chains that may contain up to 12 different sugars linked between them with up to 22 different glycosidic linkages (Mohnen 2008). RG-I domain, on the other hand, has a backbone made of a repeating unit of α(1,4)-D-GalA-α(1,2)-L-rhamnose disaccharide. Rhamnose (Rha) residues can bear oligosaccharide side chains of β(1,4)-linked D-Gal residues, or α(1,5)-linked L-Ara residues (arabinan and galactan branches in Fig. 1.2). Pectin constitutes a minor component (up to 10 % w/w) in woody plant biomass and grasses. Thus, they have less impact on biofuel production from plant biomass.

1.3.1.4 Lignin

Lignin is the result of combinatorial radical (oxidative) coupling of phenylpropanoid subunits (called *p*-coumaryl, guaiacyl, and syringyl alcohols in Fig. 1.3a). These three precursors differ in the number of methoxyl groups (-OCH$_3$) on the

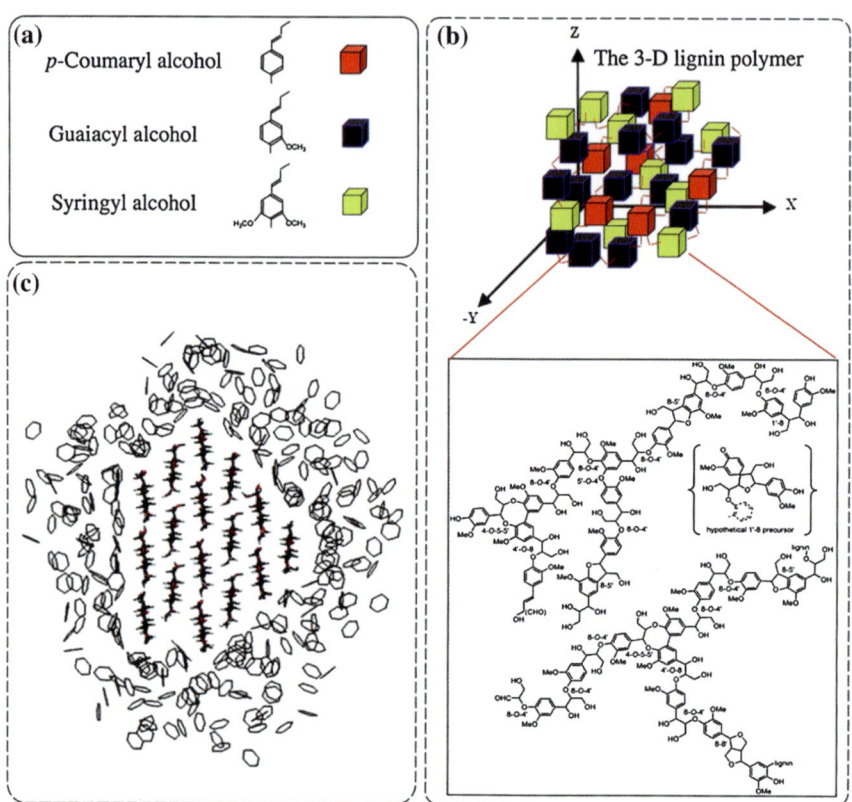

Fig. 1.3 Chemical structure and organization of lignin polymer in plant biomass. **a** Chemical structures and symbols of lignin monomers (the three monolignols *p*-coumaryl, coniferyl, and sinapyl alcohols). **b** Schematic representation of a polymerized lignin polymer in its 3-D structure, along with 8-O-4, 8-5, 8-8, 8-1, 5-5, and 4-O-5 linkages that interconnect *p*-hydroxyphenylpropanoid units in the polymer (source, DOE-Oak Ridge National Laboratory). **c** "Lignin-cellulose" complex presented as a drawing of the bilayer of 142 lignin dimers adsorbed on a cellulose microfibril surface as predicted by molecular medeling (*source* Besombes and Mazeau 2005)

aromatic ring (Fig. 1.3a). The oxidative coupling reactions, catalyzed by peroxidases or laccases (Higuchi 1985; Sterjiades et al. 1992; Ranocha et al. 2002), proceed in 3-D dimension (Fig. 1.3b), and seem to follow a high degree of selectivity (Erickson et al. 1973; He and Terashima 1989). This high degree of selectivity is dictated by a class of proteins called dirigent proteins (Davin and Lewis 2000). Lignin is considered the most abundant of all biopolymers.

Plants designed lignin as a strong glue that holds together and impermeabilizes "cellulose-hemicellulose" network (Fig. 1.3c), providing tissues with both the strength to resist pathogen attacks and the capacity to limit water loss (Rubin 2008). Therefore, lignin limits accessibility to cellulose and hemicellulose, which hinders the complete enzymatic hydrolysis of these polymers (Chen and Dixon 2007). The assembly of "cellulose-lignin" complex has been investigated through molecular modeling using lignin oligomers having up to 20 subunits. The data showed that, from the energetic stand, the association "cellulose–lignin" is preferred over "lignin–lignin" (Besombes and Mazeau 2005; Erilkssen et al. 1980). This means that during early stages of lignin deposition, lignin monomers/dimers tend to cover most cellulose microfibrils and hemicellulose surfaces available by adsorption (via hydrogen bonding to create a film on cellulose microfibril surface) before lignin self-aggregation. In later stages of lignin deposition, "lignin–lignin" complexes are formed and locked by oxidative coupling reactions catalyzed by peroxidases and/or laccases. As a result, cellulose microfibrils and hemicellulose are glued by a 3-D lignin polymer (Fig. 1.3b, c). Thus, lignin is still one of the main obstacles in the conversion of lignocellulosic plant biomass into bioethanol (Lynd et al. 1991; Hill et al. 2006). Any pretreatment that enhances the release of cellulose and hemicellulose from lignin would tremendously impact the cost of bioethanol production. Manipulation of oxidative coupling reactions in later stages of lignocellulosic formation would logically lower the recalcitrance of plant biomass. Also, identifying microbes able to metabolize lignin, as carbon source, would be of interest.

Lignin composition is cell types- and species-dependent and is affected by environmental conditions. For example, softwood lignin contains mostly guaiacyl alcohol (G-monomer), while hardwood lignin contains almost equal amounts of guaiacyl and syringyl (S-monomer) units (Boerjan et al. 2003; Ralph et al. 2004, 2008). However, grass lignin contains *p*-coumaryl (hydroxyphenyl, H-monomer) units in addition to G- and S-monomers. Because of the differences in the chemical bonds that link G-, S-, and H-monomers within lignin polymer, plant biomass can have different degree of recalcitrance. An increase in S-monomer content in lignin is believed to confer more recalcitrance to plant biomass compared to G-monomer (Li et al. 2001; Kishimoto et al. 2010). Modification of lignin composition in transgenic plants has been shown to result in significant improvements in the conversion of biomass into bioethanol (Chen and Dixon 2007).

1.3.2 Macromolecular Organization: Primary Versus Secondary Cell Walls

The polymers described in the previous section (cellulose, hemicellulose, pectin, and lignin) interact in a specific manner to form a functional and complex network, the cell wall. Each polymer has a specific physiological function, for example, cellulose microfibrils are the load-bearing component of the cell walls. These microfibrils are held together by hemicellulosic polymers (hence the name of cross-linking polysaccharides) to give additional strength to the cell wall (Hayashi and Maclachlan 1984; McCann et al. 1990; Carpita and Gibeaut 1993; Scheller and Ulvskov 2010). The "cellulose-hemicellulose" matrix is embedded

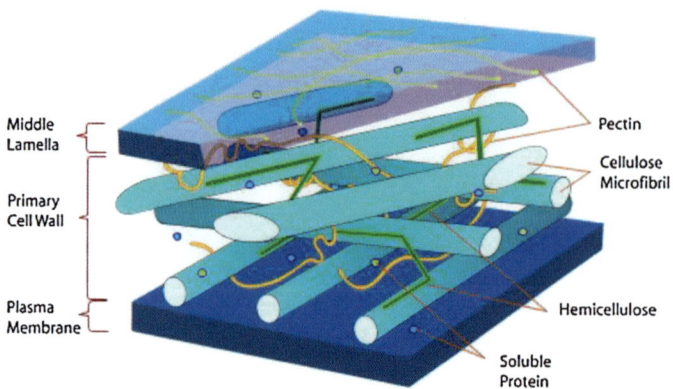

Fig. 1.4 Schematic representation of a plant cell wall model showing the interactions between cellulose, hemicellulose, and pectin within plant primary cell wall in growing tissues (*source* DOE-Oak Ridge National Laboratory). At tissue maturity, the cells of some tissues depose additional layers of cellulose microfibrils (up to 50 % of the biomass) and hemicellulose, and lock them with a 3-D lignin polymer. This additional structure is called secondary cell wall and located between primary cell wall and plasma membrane (not shown in the picture)

into another matrix made of pectin, as an adhesive. Several schematic models for how cellulose, hemicellulose, and pectin interact with each other to form architecturally functional "primary cell walls" have been proposed (Keegstra et al. 1973; McCann et al. 1990, 1992; Carpita and Gibeaut 1993). A schematic presentation of primary cell wall organization is indicated in Fig. 1.4. The primary cell walls of adjacent plant cells are glued together by middle lamella, a matrix of polymers rich in pectin (Fig. 1.4). The primary cell wall is structurally strong and yet flexible to allow its smooth remodeling during cell elongation and growth (McCann et al.1990 ;C osgrove1993).

At the end of cell elongation (transition to differentiated cells), some cells lay down additional layers of cellulose microfibrils and hemicellulose between the plasma membrane and the primary cell wall, and lock the whole polymers in a rigid, 3-D matrix of lignin. This thick and impermeable cell wall is now called the secondary cell wall (Carpita and Gibeaut 1993; McCann et al. 1990), and represents the main component of plant biomass used for biofuel production. However, despite the complexity of the structural organization of the polymers within each cell wall types, some subtle differences in their interactions (e.g., cellulose-hemicellulose or cellulose-lignin interactions) can have important impact of the costs of the processing of the biomass for biofuel production (Dien et al. 2006; Narayanaswamy et al. 2011), and forage quality (Fales and Fritz 2007). Polysaccharide contents of plant cell walls (primary and secondary cell walls) vary depending of the species, tissues, growth stage, and conditions, they all contain cellulose. However, while cellulose content (expressed in dry weight) in the primary cell walls is usually up to 30 %, it can reach up to 50 % in the secondary cell walls (Mellerowicz et al. 2001). On the other hand, hemicellulose composition shows drastic differences compared to cellulose. For example, whereas XyG is the major hemicellulose in primary cell walls of dicots and noncommelinoid monocots (These walls are also called type I walls), xylan (e.g., GAX, GX, and AX) and MLG are the most abundant hemicellulose in walls from Poales including grasses (these walls are also called type II walls) (Carpita and Gibeaut 1993). Plant biomass from feedstock is composed of secondary cell wall materials that are rich in cellulose and xylan and/or glucomannan.

1.3.3 Biosynthesisand AssemblyofPla ntC ell Walls

The process of biosynthesis and assembly of all cell wall polymers is not well understood. Currently our knowledge of the biochemistry of these processes is very limited, as we lack an understanding of the enzymology itself, secretion; coordination of gene expression, multi-enzyme complex assembly/organization; and this list is not exhaustive. Ultimately, if we want to be able to design/engineer plants for biofuel production (biotechnologically domesticate), it will be necessary to understand how plants synthesize and assemble these polymers into structurally functional cell walls. If one considers the number of different glycosidic linkages within and between plant cell wall polysaccharides, numerous enzymes would be required for wall elaboration. Only about a dozen of such enzymes have been characterized and their biochemical functions identified, in addition to at least 12 genes in xylan biosynthesis, called *irregular xylem (IRX)* for which their exact biochemical functions remain elusive (Edwards et al. 1999; Perrin et al. 1999; Faik et al. 2000, 2002; Burton et al. 2006; Sterling et al. 2006; Eglund et al. 2006; Dhugga et al. 2004; Cocuron et al. 2007; Persson et al. 2007; Doblin et al. 2009; Wu et al. 2009; 2010a, b; Jensen et al. 2011). In general, the biosynthetic enzymes can be classified into two broad groups: the enzymes that make the backbones of the polymers, called synthases, and those that decorate these backbones, called *glycosyltransferases (GTs)*. Although both types of enzymes use activated sugar residues as donor molecules, they have completely different mechanisms. Synthases are usually processive enzymes, adding sugars repeatedly to make the backbone of a polysaccharide; however, GTs add one sugar residue at a time onto a specific position of a specific acceptor. The carbohydrate-active enzymes (CAZy, http://afmb.cnrs-mrs.fr/CAZY/) database has classified these putative synthases and GTs into families referenced as "GT" and numbered from 1 to 94. According to various biochemical and structural studies, the Golgi apparatus is the intracellular site for the biosynthesis of noncellulosic polysaccharides, with callose being the exception. These studies include ultrastructural work, and work showing polysaccharide biosynthetic enzymes and activities associated with fractions enriched in Golgi vesicles.

 To gain insights into this assembly process, *Gluconacetobacter xylinus* (former *Acetobacter aceti subsp. xylinum*) was used as a system to investigate "cellu-lose-hemicellulose" network assembly in vitro. The idea is to grow *G. xylinus* in presence of hemicellulosic polymers and analyze the crystalline structure of newly synthesized cellulose microfibrils. In absence of hemicellulose, *G. xylinus* synthesized highly ordered cellulose microfibrils (crystalline structure), but in presence of hemicellulosic polymers the bacterium produced cellulose microfibrils with altered shape, similar to cellulose microfibrils synthesized by plants (Atalla et al. 1993; Whitney et al. 1995, 1998). Interestingly, these type of experiments showed that cross-linking with XyG has resulted in a dramatic weakening of cellulose microfibrils and greater extensibility of "cellulose-XyG" matrix (Whitney et al. 1999, 2006; Chaliaud et al. 2002). Similar system was used to investigate the interactions between cellulose microfibrils and lignin-like polymers (Touzel et al. 2003).

1.4 DomesticationofPlants f orB iofuel: Yield, Recalcitrance, and Carbon Sequestration

In the last decades, genomics and biotechnology have impacted plant improvement and will most likely result in domestication of plants for biofuel production. Considering that the world's energy consumption will increase by 57 % in the next 20 years (EIA 2007; Li et al. 2001), and the world population is expected to reach up to 10.5 billion by 2050 (http://www.worldometers.infor/world-population), developing a dedicated feedstock crop for biofuels production not only must take in account these two variables, and avoid "energy-food" competition, but must also answer the following goals: *Goal 1*: The plant system should allow efficient use of the biomass (gallon/ton); *Goal 2*: the plant system must allow efficient use of land (ton/acre); and *Goal 3*: the plant system must be carbon neutral (Fig. 1.5). So a plant (or a plant system) that answers all these needs would be a good fit for biofuelproduc tion.

1.4.1 Efficient Use of the Biomass (Goal 1)

The efficient use of the biomass can be achieved through several ways. For example, crops with less recalcitrant cell walls generate biomass that is easy to deconstruct without the need of expensive pretreatment steps. Efficient use of plant biomass can also be achieved through metabolic engineering of a plant to make it produce new components with low nutrient input (less fertilizer and water use), in addition to lignocellulosic biomass. Alternatively, lignocellulosic biomass can also be optimized for the production of chemicals of interests such as furans and acetic acid (Binder and Raines 2009; Hsieh 2009).

Crops with enhanced biotic (pests and diseases) resistance, and abiotic (drought, cold, salt) tolerance are also desired, as these crops would grow on nonarable land and produce high biomass yield. For example, improving biotic resistance, and abiotic tolerance of any promising C4-type plants feedstocks that have efficient carbon

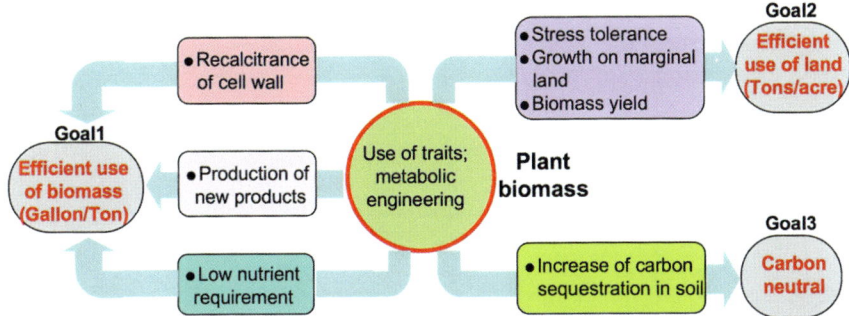

Fig. 1.5 Domestication of plant biomass for biofuels. This process would either use plant traits or metabolically engineer current plant biomass to achieve efficient use of land (tons per acre), efficient use of biomass (gallon per acre), and carbon sequestration. The targets are recalcitrance of cell wall, growth in marginal land and/or low nutrient requirement, biomass yield, stress tolerance, production of new chemicals, and carbon neutrality

fixation, water and nitrogen use (such as *Miscanthus* and switchgrass), would have an impact on biomass research field (Fig. 1.5).

1.4.2 Efficient Use of the Land (Goal 2)

In this regarding, one should consider the energy content of the biofuel produced from renewable biomass (bioethanol versus biodiesel). For example, a liter of bioethanol has only two-thirds of energy density of biodiesel, making the cars that run on E85 (85 % ethanol and 15 % diesel) have ~30 % lower gas mileage (Regalbuto 2009). Because biodiesel is a more energy-rich biofuel, it is also suited for fueling aircraft and heavy trucks (Service 2011). Biodiesel can be produced by reacting vegetable oils (typically triacylglycerides) from soybean, sunflower, or oil palm with an alcohol, which results in the production of fatty acid methyl esters (Hu et al. 2008). Thus, there is an increasing interest in the fast-growing algae that can produce up to 40 % of its weight in oil, 6 times more than the most producing plants such as oil palm (Fig. 1.6a) (Posten 2009; Stephenson et al. 2011; Araujo et al. 2011). However, while the fast-growing algae can be a good choice in many regards (e.g. use land unsuited for agriculture, only small area is needed), until now, the costs of the extraction and processing of the oil are still high ($2.25 per liter),a ndw aterus ageis high(Se rvice2011).

An alternative way to achieve an efficiently use of land is to select feedstock crops that are easy to engineer for better growth on nonarable land with minimal nutrient inputs to avoid competition for food crops, and lower the production costs. It would be even better if this selected feedstock crop has both enhanced carbon assimilation (e.g., C4 photosynthetic plants), and sustainable rapid growth cycle (e.g., crops with multiple harvests per season). Currently, *Miscanthus*

Fig. 1.6 The two most promising feedstock plants for biofuels currently under investigation. **a** Oil palm, *Elaeis guineensis* Jacq., a tropical tree species that produces oil-rich fruit resembling avocados, and yields up to 10.6 tons of oil per hectare per year. The oil has similar composition to soy oil and well suited for biodiesel usages (*credit* Somerville c., Stanford University; *source* DOE-Oak Ridge National Laboratory). **b** *Miscanthus* is a promising feedstock crop that is currently under investigation in the United States (*Credit* Long S., University of Illinois, *source* DOE-Oak Ridge National Laboratory)

(Fig. 1.6b) appears to have these advantages, as it is a C4 plant exhibiting greater photosynthetic efficiency, and requires less water and mineral nutrients compared to corn or other grasses. In addition, *Miscanthus* can grow up to 3.5 m in height, which would produce an annual 10 tons biomass per acre, compared to 7.6 and 4 tons biomass/acre for corn and wood timber, respectively (U.S. DOE. 2006). In terms of ethanol production, *Miscanthus* can produce up to 165 gallons ethanol per ton of biomass compared to 95 gallons per ton for corn.

1.4.3 Efficient Reduction of Carbon Release (Goal 3)

Improved capacity to sequester assimilated carbon in organs below ground is important for negative carbon cropping. In this regard, perennial grass species (e.g., switchgrass and *Miscanthus*) appear to have superior capacity in sequestering captured carbon in their roots (Tilman et al. 2006). Therefore, these feedstock crops would have less carbon emission, which would limit greenhouse gas effects (GHG) due to burning of fossil fuels and industrial processes (Alder et al. 2007).

In summary, although *Miscanthus* has attractive qualities to answer the needs of biofuel industry, it does not have all the desired traits, and most likely a mix of species that are adapted to various climatic regions will be engineered to maximize

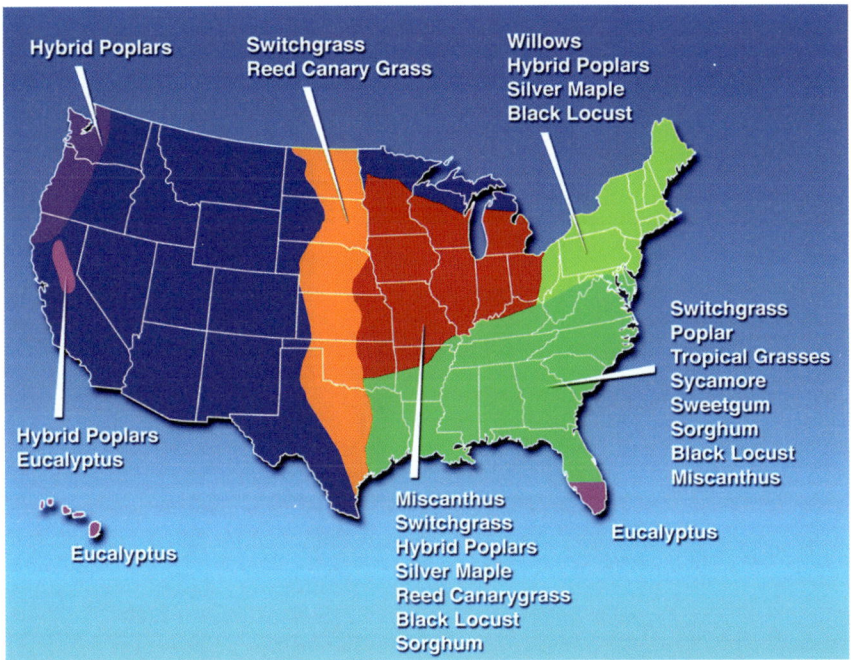

Fig. 1.7 Feedstock crops currently under investigation in the United States along with their geographic distribution. These crops have various ecological habitats and optimizing several of them would be required to cover U.S. liquid fuel needs (*source* DOE·Oak Ridge National Laboratory)

biofuel production (Fig. 1.7). Screening for the best fit should continue, and in this regard I would like to cite what Thomas Jefferson already said in his letter to William Drayton in 1786, *We are probably far from possessing, as yet, all the articles of culture [crops] for which nature has fitted our country. To find out these, will require an abundance of unsuccessful experiments. But if, in a multitude of these, we make one or two useful acquisitions, it repays our trouble.* This is still valid now, and with a scientific progress in genomics, metabolic engineering, and screening nature's biodiversity, we should be able to achieve this goal within a reasonable amount of time.

1.5 Plant Biomass Conversion: the Good, the Bad, and the Difficult

The general scheme of bioethanol production from biomass involves three steps (Fig. 1.8): step 1: *biomass pretreatment* to deconstruct cell walls into accessible cellulose ("the good"), hemicelluloses ("the difficult"), and eliminating other polymers such as lignin ("the bad"); step 2: *hydrolysis/saccharification* of the pretreated

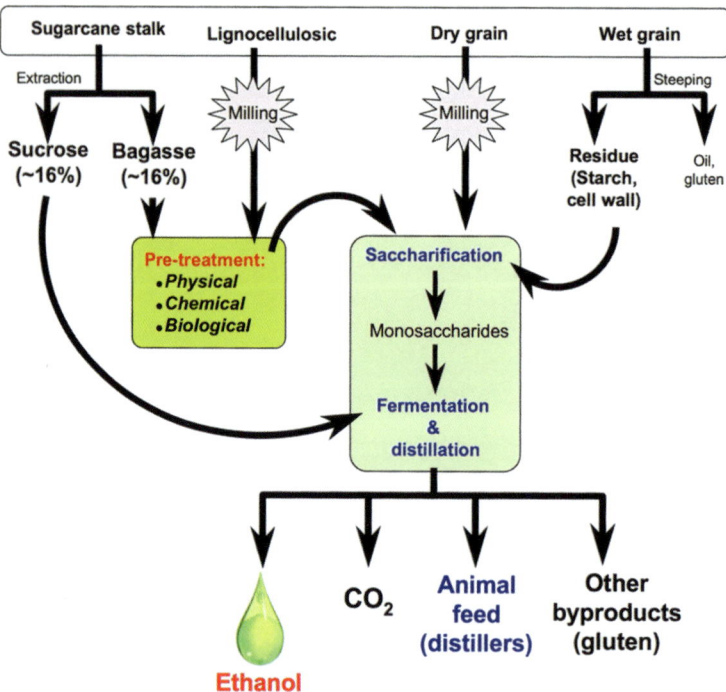

Fig. 1.8 General steps of biochemical conversion of various types of biomass into ethanol and other byproducts. The starting biomass material can be lignocellulosic, bagasse (rich in recalcitrant fibers) or grains (rich in starch and/or oil). Pretreatment step, however, is required only for recalcitrant plant biomasses such as lignocellulosic and bagasse. After this pretreatment step, the following steps in ethanol production process are common for all biomasses, and consist in saccharification to produce monosaccharides, which are then fermented to produce ethanol as well as other coproducts (e.g., distillers, gluten). Distillers are usually added to animal diets (beef, swine, and poultry) up to 20 %(w/w), which increase the livestock industry

biomass; and step 3: *fermentation* of the monosaccharides (6- and 5-carbon sugars) into ethanol. Each of these steps has difficulties that add costs to bioethanol production.

1.5.1 PretreatmentandSac charification Basics

The efficiency of the conversion of plant biomass to biofuels (particularly bioethanol) depends on how much efforts (and costs) are required to overcome cell wall recalcitrance. Currently, pretreatment step, which allows accessibility of the enzymes to cellulose and hemicellulose, is the most costly step in the whole process (Mosier et al. 2005). There is a positive correlation between cell walls recalcitrance and their lignin content/type. Plant biomasses that are rich in lignin require harsh and costly pretreatment to free cellulose and hemicellulose. Moreover, the degree of crystalline structure and bundling of cellulose microfibrils can also limit the accessibility of enzymes to dissociated glucan chains. For example, woody biomass (especially softwood) is very recalcitrant because is made of secondary cell walls that are rich in lignin (up to 40 %) and their cellulose microfibrils are highly crystalline and form thick bundles (up to 68 nm in diameter) (Ding and Himmel 2006; Kennedy et al. 2007; Xu et al. 2007). Lignin is linked to carbohydrates (cellulose and hemicellulose) through benzyl ester, benzyl ether, and glycosidic linkages. The benzyl ester linkages can be broken by alkali treatments, but benzyl ether and glycosidic linkages are resistant to alkali hydrolysis and require harsher conditions. Currently, pretreatment strategies used in cellulosic conversion include a wide range of physico-chemical reactions such as temperature (90–250 °C), pH, and fiber explosion (stem, ammonia, and CO_2) (Mosier et al. 2005; Narayanaswamy et al. 2011). However, some of these pretreatment methods have additional impact on the efficiency of the enzymes used in saccharification, and can also inhibit the growth of the fermenting microorganisms (Saha and Cotta 2010; Rosgaard et al. 2007). Therefore, extra costs are usually encumbered to remediate the effect of pretreatment methods on the enzymes.

Saccharification is the second most costly step in biofuel production process due to the price of the enzymes. So far, pretreatment with high temperature is the most compatible with subsequent use of enzymes without additional adjustment steps, but it usually involves high-energy usage and degradation of labile/unstable sugars. To overcome this issue, Narayanaswamy et al. (2011) recently proposed super critical CO_2 pretreatment to limit sugar loss and lower energy usage.

The knowledge learned from structural studies on plant cell walls suggests that no universal pretreatment method is optimal for every plant biomass. Thus, pretreatments should be designed on the basis of this knowledge. However, pretreatment techniques can be grouped into three distinct categories: *physical, chemical*, and *biological pretreatments*. These categories can be combined, as two or more pretreatment techniques can be incorporated for efficient result. Any pretreatment combination that allows (i) minimal use of energy input, catalyst recycling, value-added chemicals, and low impact on downstream processes, (ii) recovery of sugars from hemicellulose portion of biomass, and (iii) limiting the production of inhibitors, is highly desired.

1.5.1.1 PhysicalPr etreatments

These pretreatments involve chemical-free methods such as steam water explosion (autohydrolysis), liquid hot water, and mechanical grinding (e.g., milling). Currently autohydrolysis is the most cost-effective pretreatment used in pilot scale (Hsu 1996). It consists of inducing explosive decompression by rapid heating of the biomass (up to 160–260 °C) under high-pressure (saturated steam of water at 0.69--4.83 MPa) for a short time before a sudden release of pressure (McMillan 1994; Sun and Cheng 2002). While heat and pressure induce hemicellulose solubilization (and also its partial hydrolysis by organic acids), pressure decompression produces dissociation of cellulose microfibrils and "cellulose-lignin" complexes (Michalowicz et al. 1991), which allows access of enzymes to glucan chains. In liquid hot water pretreatment (opposed to steam water), plant biomass is basically cooked under pressures that maintain water in the liquid state (Rogalinski et al. 2008). This method can release up to 80 % of hemicellulose, but only certain amount of lignin was released from the biomass.

However, steam explosion and water cooking methods may not be very effective on lignocellulosic biomass from grasses such switchgrass (Alizadeh et al. 2005) and *Miscanthus* (Brosse et al. 2009), most likely because of the high strength of cellulose microfibrils in these biomasses, and the release of furfurals. These organic compounds are result of the loss of three water molecules (dehydration) from C5-carbon sugars (e.g., Xyl and Ara) to form aldehydes. Furfurals can be toxic to certain fermenting microorganisms.

1.5.1.2 ChemicalPr etreatments

They involve the use of chemicals to help disintegrate the biomass before saccharification. Most methods utilize acids (e.g., SO_2, H_2SO_4, and other acids), alkali (e.g., calcium hydroxide, potassium hydroxide, ammonia, hydrogen peroxide, and others), or organic solvent (e.g., ethanol and methanol) (Zhang et al. 2007). These methods are, in a sense, similar to steam water explosion. For example, the biomass is impregnated in SO_2 or H_2SO_4 solutions, and then subjected to steam explosion as described in the previous section (De Bari et al. 2007). Comparatively, alkali pretreatments use in general lower temperatures and pressures than pretreatments with acid solutions. Also, CO_2 appears to be less effective than SO_2 and H_2SO_4 (Mackie et al. 1985), but SO_2 has negative effects on health and environment. However, recent study using super critical CO_2 showed promises for grass biomass pretreatment (Narayanaswamy et al. 2011).

The other chemical pretreatment widely used is ammonia fiber/freeze explosion (AFEX). This method uses hot liquid ammonia under high-pressure that is rapidly released to induce explosion. The main advantage of this method is that it removes most lignin and at the same time dissociates cellulose microfibrils without major loss of hemicellulose. This method has been successfully applied to lignocellulosic biomass from grasses such as switchgrass and *Miscanthus*. Another delignification process, called organosolv, consists in using an organic solvent such as ethanol and methanol (in presence or absence of acid or alkali chemicals) to break lignin-carbohydrates linkages (Pan et al. 2007; Taherzadeh and Karimi 2008). Although organosolv is an efficient method that improves biomass conversion to ethanol, its use is costly due to their high volatility, which requires expensive high-pressure equipment (Aziz and Sarkanen 1989). Thus, this method is not as attractive for industrial scale production as the other pretreatment methods.

1.5.1.3 BiologicalPr etreatments

The physical and chemical pretreatments described above require subsequent use of enzymes to degrade cellulose and hemicellulose before fermentation. In biological pretreatments, microorganisms are used to directly degrade lignocellulosic biomass and initiate fermentation process at the same time. As a result, lactate, ethanol, acetate, hydrogen, and CO_2 are directly produced (Kurakake et al. 2007; Lee et al. 2007). These microorganisms are able to degrade solid biomass because they secrete hydrolases that bind to insoluble cell wall polymers through carbohydrate binding domains. They are able to synthesize large multi-enzyme complexes called cellulosomes that can bind and degrade cellulose and hemicellulose (Bayer et al. 1998; Lynd et al. 2002). Usually, these microorganisms are thermophilic and can still grow in temperatures higher than 50–60 °C. For example, *Clostridium thermocellum*, a cellulosome producing bacterium, has an optimal growth around 60 °C (Bayer et al. 1998). Other anaerobic thermophiles such as *Anaerocellum thermophilum* strain Z-1320 (Blumer-Schuette et al. 2008; Svetlichnyi et al. 1990) and strain DSM 6725 (Yang et al. 2009) can grow in 75 °C. Because of these high temperatures required for bacterial growth, there is less bacterial contamination during fermentation, a problem that is not desired in traditional fermentation process (Skinner and Leathers 2004). Currently efforts are directed toward the identification of microorganisms that can efficiently degrade or modify lignin. White-rot fungi (Basidiomycetes) appear to be promising candidates for biofuels production from solid biomass (Hwang et al. 2008; Kuhar et al. 2008). The main disadvantage is that most of these fungi use the energy generated by polysaccharide fermentation to degrade lignin, which results in a lower fermentation yield. It would be necessary to develop strains that use lignin itself (instead of polysaccharides) to generate the energy needed for lignin degradation. Such strains are valuable even if they are not efficient in fermenting sugars, as they can be used in combination with traditional fermentation with limited loss of sugars. There are many other advantages in developing biological pretreatment for biofuels. First, some of these microorganisms have their genome already sequenced (Kataeva et al. 2009), which would open the door to identifying "super-active" enzymes that can degrade insoluble cellulose and hemicellulose in the biomass before carrying out traditional fermentation. With the progress in genomics and next-generation sequencing, it is possible now to screen for other thermophilic bacteria, and easily sequence their genomes.

1.5.2 FermentationB asics

Efficient fermentation of sugar mixtures released from pretreated and saccharified lignocellulosic biomass is crucial for successful large-scale bioethanol production. Until recently, most researchers focused on the conversion of the cellulosic part (composed of Glc, a 6-carbon sugar) of the biomass and neglected the hemicellulose portion (composed mainly of Xyl and Ara, both are 5-carbon sugars, called pentoses). Pentoses are not or very difficult to ferment to ethanol by mostly used microorganisms in traditional fermentation process. This wasted biomass represents up to 40 % (w/w) of the total sugars (Fig. 1.1). Currently, *Saccharomyces cerevisiae* is the microorganism mostly used in large-scale fermentation of Glc from sugarcane and starch. However, this microorganism is not suited for fermenting sugars from plant biomass that are rich in pentoses (van Maris et al. 2006). The first microbes (e.g. *Zymomonas mobilis*) that could ferment these pentoses were discovered in the 1970s, which allowed engineering *Escherichia coli* and *S. cerevisiae* for the conversion of hexoses and pentoses (Ohta et al. 1991). Despite this progress, cofermenting simultaneously Glc and pentoses in a mixture is still far from optimal for bioethanol production, on a commercial scale. The bottleneck comes from "glucose repression" effect, as most fermenting yeast do not metabolize other sugars when Glc is present in the mixture, until Glc is depleted (Trumbly 1992; Kim et al. 2012). "Glucose repression" effect inhibits the uptake and metabolism of other sugars (Fig. 1.9). Therefore, to improve efficient use of plant biomass in bioethanol production, efforts have been focusing on engineering yeast *S. cerevisiae* for cofermentation of Glc and pentoses (Kim et al. 2012). The first step was to engineer *S. cerevisiae* into Xyl-fermenting yeast by introducing the genes that encode for enzymes involved in the conversion of Xyl to pyruvate. This pathway consists of three main genes (red in Fig. 1.9b): *Xyl reductase* (*XYL1* or *XR*), *xylitol dehydrogenase* (*XDH*), and *xylulokinase* (*XKS*). These genes were isolated from Xyl-fermenting fungi such as *Pichia stipitis*, *Candida tenuis*, and *Spathaspora passalidarum* (Ho 1998; Eliasson et al. 2000). The other alternative is to engineer *S. cerevisae* strains that overexpress both *xylose isomerase* (*xylA*) and *xylulokinase* (*XKS*) genes allowing an efficient fermentation of Xyl (red in Fig. 1.9b) (Walfridsson et al. 1996; Brat et al. 2009;K arhumaae ta l.2005).

Currently, ethanol industry has given priority to metabolically engineer these Xyl-fermenting *S. cerevisiae* strains into strains that can coferment Glc and Xyl. However, "glucose repression" effect is still a big handicap in this process. To overcome this issue, several approaches have been employed to either (i) generate *S. cerevisiae* strains that are defective in glucose-sensing or signaling pathway that is involved in "glucose-repression" effect (Kraakman et al. 1999; Versele et al. 2004; Hector et al. 2008). In these strains, the uptake and metabolism of other sugars in presence of Glc would be eliminated, or (ii) generate *S. cerevisiae* strains overexpressing transporters for cellobiose and Xyl (*S. cerevisiae* lacks Xyl-specific transporter), along with β-glucosidase (breaks down cellobiose to two Glc residues) activity that would eliminate Glc accumulation in the medium and stimulate the uptake of Xyl. *S. cerevisiae* strains overexpressing β-glucosidase enzyme and cellobiose transporter improved ethanol production yield, as Glc formed from cellobiose is rapidly assimilated, which reduces Glc repression effect. Cellobiose does not exert any repression effect on the metabolism of other sugars. Thus, these engineered *S. cerevisiae* strains allowed efficient cofermentation of Glc and Xyl and an increase in ethanol production (Nakamura, et al. 2008). It is clear that

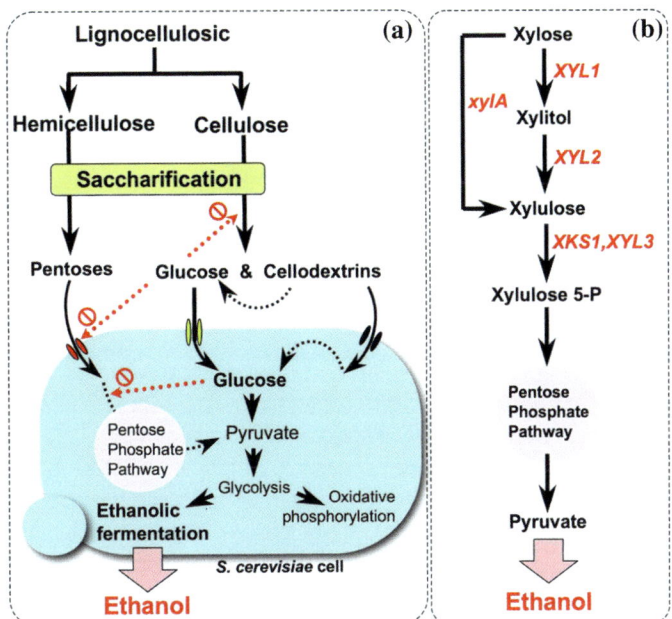

Fig. 1.9 Schematic presentation of yeast *(Saccharomyces cerevisiae)* fermentation metabolic pathway of sugars derived from plant biomass. **a** Glucose, cellodextrins, and pentoses produced from hemicellulose and cellulose are transported inside of yeast cells through transporter proteins to be metabolized via pentose phosphate pathway and/or glycolysis before either conversion to ethanol via ethanolic fermentation (anaerobic conditions), or oxidative phosphorylation in the mitochondria (aerobic conditions). "Glucoes-repression" effect is illustrated by the inhibition (X sign) of pentose uptake and metabolism, and the inhibition of complete cellulose degradation by enzymes. **b** Metabolic engineering of xylose-fermenting yeast to enhance simultaneous fermentation of hexose and pentose sugars. Heterologous expression of genes encoding enzymes involved in the conversion of xylose to pyruvate, namely xylose reductase (XYLl), xylitol dehydrogenase (XYL2), xylose isomerase (xyIA), and xylulokinase (XKS I, XYL3), allowed the generation of engineered S. *cerevisiae* strains that are able to co-ferment xylose and glucose present in mixtures

metabolic engineering of *S. cerevisiae* is a promising strategy and is currently given a priority in the conversion of lignocellulosic biomass to bioethanol. For example, improving the activity (Km, Vm) of the proteins listed above by protein engineering would allow enhanced Xyl metabolism in yeast (Kim et al. 2012).

Progress in genomics and high-throughput screening methods could result in the discovery of new genes or gene variations, such as Xyl transporters that are not inhibited by Glc, would improve cofermentation process and bioethanol production (Ni et al. 2007; Liu and Hu 2011; Peng et al. 2012). However, despite all these intensive efforts and until now, any engineering that strongly reduces Glc fermentation efficiency, also negatively impacts the assimilation and fermentation of Xyl. Although cofermentation of cellulose and xylan through overexpression of glucosidase and xylosidase at the surface of *S. cerevisiae* is now possible, it still needs optimization for a large industrial scale. Alternatively, engineering grass plants (particularly C4 plants) with less pentose content (e.g., xylan and arabinan) in their cell walls, would improve biomass conversion.

During fermentation, ethanol is recovered from the mixture by distillation followed by adsorption or filtration. The residual materials (e.g., lignin, ash, enzyme, microorganism debris, nonfermented cellulose, and hemicellulose) are recovered as solid for burning or can be converted to various value-added by-products (Fig. 1.8).

1.6 ConcludingR emarks

Plant biomass holds great promise as raw material for the biofuel renewable energy sector, and also for textile and food industries. Thus, in term of fossil energy ratio (FER, also called energy balance), which is an indication of whether the fuel is renewable and is defined as the energy delivered to customer divided by fossil energy used, cellulosic feedstock ethanol is 2 times better than biodiesel (from soybean oil), 3–4 times better than corn ethanol, and more than 10 times better than electricity (Fig. 1.10, Sheehan et al. 1998, 2003). Thus, while electricity is almost nonrenewable, lignocellulosic ethanol appears to provide the best sustainability scenario. In addition, the fact that energy is provided from lignocellulosic biomass itself results in a reduction of CO_2re leasebyupto70%.

The main current recurring debate is also dominated by "food or fuel" balance. The emerging picture from this debate is that there is a need for domestication of several feedstock crops, because none of the current available feedstock crops have all the requirements to balance our food and fuel needs. With the current scientific progress in genomics, it is possible to metabolically engineer the most promising plant candidates (*Miscanthus*, oil palm, and others) to fit our needs.

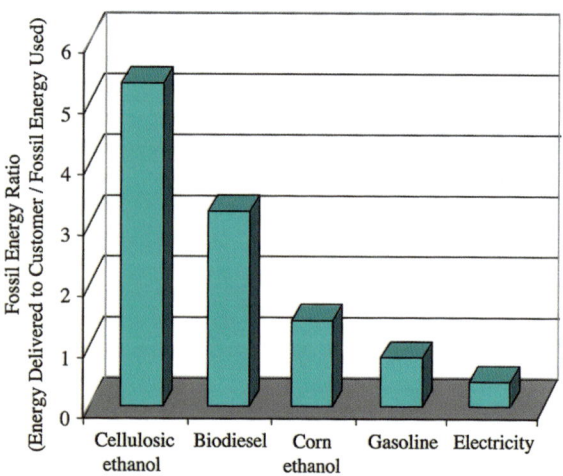

Fig. 1.10 Comparison of the ratio of the energy output of the final biofuel product to the fossil energy required to produce it for lignocellulosic ethanol, biodiesel, corn ethanol, gasoline, and electricity. Adapted from Sheehan et al. 1998 and 2003)

However, efforts to develop high-throughput screening of plant diversity and adaptation should continue, as they may bring new discoveries to biofuel field. Meanwhile, close consideration of the variations in plant cell wall structure must be taken in consideration when developing cost-effective pretreatment methods for second generation of biofuels. Plant cell walls are complex and sophisticated networks made of cellulose, hemicellulose, and lignin. Variations in the architectural interactions, even subtle ones, between these polymers within these cell walls may significantly contribute to recalcitrance of plant biomass. Many pretreatment techniques currently used showed varying efficiency in loosening different parts of cell walls in plant biomass. For example, steam explosion achieves a high hemicellulose yield and low lignin solubility while ammonia fiber explosion pretreatment is particularly effective at delignification. Because there are many different kinds of biomass, no single pretreatment method is expected to be the preferred universal choice. An in-depth understanding of various cell wall structures will help in the selection process. Continued improvements in pretreatment methods are desired to improve the economics of the lignocellulosic fuel production.

References

Alder PR, Del Grosso SJ, Parton WJ (2007) Life cycle assessment of net greenhouse gas flux for bioenergyc roppings ystems.Ec ol Appl17: 675–691

Alizadeh H, Teymouri F, Gilbert TI, Dale BE (2005) Pretreatment of switchgrass by ammonia fibere xplosion(AFEX). ApplB iochemB iotechnol121: 1133–1141

Araujo GS, Matos LJ, Gonçalves LR, Fernandes FA, Farias WR (2011) Bioprospecting for oil producing microalgal strains: evaluation of oil and biomass production for ten microalgal strains.B ioresour Technol102: 5248–5250

Atalla RH, Hackney JM, Uhlin I, Thompson NS (1993) Hemicelluloses as structure regulators in the aggregation of native cellulose. Int J Biol Macromol 15:109–112

AzizS ,Sa rkanenK (1989)O rganosolvpul ping-ar eview. TappiJ 7 2:169–175

Bayer EA, Shimon LJ, Shoham Y, Lamed R (1998) Cellulosomes-structure and ultrastructure. J StructB iol124:221–234

Besombes S, Mazeau K (2005) The cellulose/lignin assembly assessed by molecular modeling. Part 2: seeking for evidence of organization of lignin molecules at the interface with cellulose. Plant Physiol Biochem 43:277–286

Binder B, Raines RT (2009) Simple chemical transformation of lignocellulosic biomass into furans for fuels and chemicals. J Am Chem Soc 131:1979–1985

Blumer-Schuette SE, Kataeva I, Westpheling J, Adams MW, Kelly RM (2008) Extremely thermophilic microorganisms for biomass conversion: status and prospects. Curr Opin Biotechnol19:21 0–217

Boerjan W,R alphJ ,B aucherM (2003)Li gninbi osynthesis. AnnuR evP lantB iol54: 519–546

Brat D, Boles E, Wiedemann B (2009) Functional expression of a bacterial xylose isomerase in *Saccharomyces cerevisiae*. ApplEn vironM icrobiol75: 2304–2311

Brosse N, Sannigrahi P, Ragauskas A (2009) Pretreatment of *Miscanthus* x *giganteus* using the ethanol organosolv process for ethanol production. Ind Eng Chem Res 48:8328–8334

Brown RM Jr, Saxena IM, Kudlicka K (1996) Cellulose biosynthesis in higher plants. Trends PlantSc i1:149–156

Buckeridge MS (2010) Seed cell wall storage polysaccharides: models to understand cell wall biosynthesis and degradation. Plant Physiol 154:1017–1023

Burton RA, Fincher GB (2009) (1,3;1,4)-beta-D-glucans in cell walls of the *poaceae*, lower plants, and fungi: a tale of two linkages. Mol Plant 2:873–882

Burton RA, Wilson SM, Hrmova M, Harvey AJ, Shirley NJ, Medhurst A, Stone BA, Newbigin EJ, Bacic A, Fincher GB (2006) Cellulose synthase-like CslF genes mediate the synthesis of cell wall (1,3;1,4)-β-D-glucans. Science 311:1940–1942

Campbell CJ (2006) The Rimini protocol an oil depletion protocol: heading off economic chaos and political conflict during the second half of the age of oil. Energy Policy 34:1319–1325

Carpita NC, Gibeaut DM (1993) Structural models of primary cell walls in flowering plants: consistency of molecular structure with the physical properties of the walls during growth. Plant J3:1–30

Carvalheiro F, Silva-Fernandes T, Duarte LC, Gírio FM (2009) Wheat straw autohydrolysis: process optimization and products characterization. Appl Biochem Biotechnol 153:84–93

Chaliaud E, Burrows KM, Jeronimidis G, Gidley MJ (2002) Mechanical properties of primary plant cell wall analogues. Planta 215:989–996

Chen F, Dixon RA (2007) Lignin modification improves fermentable sugar yields for biofuel production. Nat Biotechnol 25:759–761

Chundawat SPS, Venkatesh B, Dale BE (2007) Effect of particle size based separation of milled corn stover on AFEX pretreatment and enzymatic digestibility. Biotechnol Bioeng 96:219–231

Cocuron J-C, Lerouxel O, Drakakaki G, Alonso AP, Liepman AH, Keegstra K, Raikhel NV, Wilkerson CG (2007) A gene from the cellulose synthase-like C family encodes a β-1,4 glucan synthase. Proc Natl AcadSc iU S A140: 8550–8555

CosgroveD J(1993)H owd op lantc ellw allse xtend?P lantP hysiol1 02:1–6

DavinL B,L ewisN G(2000)D irigentp roteinsa ndd irigents itese xplaint hem ystery of specificity of radical precursor coupling in lignan and lignin biosynthesis. Plant Physiol 123:453–462

De Bari I, Nanna F, Braccio G (2007) SO_2-catalyzed steam fractionation of aspen chips for bioethanol production: optimization of the catalyst impregnation. Ind Eng Chem Res 46:7711–7720

Deguchi S, Tsujii K, Horikoshi K (2006) Cooking cellulose in hot and compressed water. Chem Commun31:3293–3295

Dhugga KS, Barreiro R, Whitten B, Stecca K, Hazebroek J, Randhawa GS, Dolan M, Kinney A, Tomes D, Nichols S, Anderson P (2004) Guar seed β-mannan synthase is a member of the cellulose synthase super gene family. Science 303:363–366

Dien BS, Jung HG, Vogel KP, Casler MD, Lamb JFS, Iten L, Mitchell RB, Sarath G (2006) Chemical composition and response to dilute-acid pretreatment and enzymatic saccharification of alfalfa, reed canarygrass, and a switchgrass. Biomass Bioenergy30: 880–891

Ding SY, Himmel ME (2006) The maize primary cell wall microfibril: a new model derived from direct visualization. J Agric Food Chem 54:597–606

Doblin MS, Pettolino FA, Wilson SM, Campbell R, Burton RA, Fincher GB, Newbigin E, Bacic A (2009) A barley cellulose synthase-like CSLH gene mediates (1,3;1,4)-beta-D-glucan synthesis in transgenic Arabidopsis. Proc Natl Acad Sci U S A 106:5996–6001

DureeP(2007) Biobutanol:a na ttractivebi ofuel.B iotechnolJ 2: 1525–1534

Ebringerova A, Heinze T (2000) Xylan and xylan derivatives—biopolymers with valuable properties, 1—naturally occurring xylans structures, procedures and properties. Macromol RapidC ommun2 1:542–556

Edwards ME, Dickson CA, Chengappa S, Sidebottom C, Gidley MJ, Reid JSG (1999) Molecular characterization of a membrane-bound galactosyltransferase of plant cell wall matrix polysaccharide biosynthesis. Plant J 19:691–697

Eglund J, Peterson BL, Motawia MS, Damager I, Faik A, Olsen CE, Ishii T, Clausen H, Ulvskov P, Geshi N (2006) Biosynthesis of pectic rhamnogalacturonan II: molecular cloning and characterization of golgi-localized alpha(1, 3) xylosyltransferases encoded by RGXT1 and RGXT2 genes of Arabidopsis thaliana. Plant Cell 18:2593–2607

EIA (2007) International Energy Outlook 2007 with projection to 2030. http://www.eia.doe. gov///oiaf/aeo/index.html. Accessed12M arch2009

Eliasson A, Christensson C, Wahlbom CF, Hahn-Hägerdal B (2000) Anaerobic xylose fermentation by recombinant *Saccharomyces cerevisiae* carrying XYL1, XYL2, and XKS1 in mineral medium chemostat cultures. Appl Environ Microbiol 66:3381–3386

EricksonM ,M ikscheG E,S omfaiI (1973)H olz-a ndf orschung271: 13–119

Erilkssen O, Goring DAI, Lindgren BO (1980) Structural studies on chemical bonds between ligninsa ndc arbohydratesi ns prucew ood. WoodS ci Technol1 4:267–279

Faik A(2010)X ylanb iosynthesis:n ewsf romt heg rass.P lantP hysiol1 53:396–402

Faik A, Bar-Peled M, Derocher E, Zeng W, Perrin RM, Wilkerson C, Raikhel NV, Keegstra K (2000) Biochemical characterization and molecular cloning of an α(1,2)Fucosyltransferase that catalyzes the last step of cell wall xyloglucan biosynthesis in pea. J Biol Chem 275: 15082–15089

Faik A, Price NJ, Raikhel NV, Keegstra K (2002) An Arabidopsis gene encoding an α-xylosyltransferase involved in xyloglucan biosynthesis. Proc Natl Acad Sci U S A 99:7797–7802

Fales S, Fritz JO (2007) Factors affecting forage quality. In: Barnes RF, Nelson CJ, Moore KJ, Collins M, (eds) Forages, the science of grassland agriculture, vol 2. Blackwell Publishing, Ames, pp 569–580

Fry SC (1986) Cross-linking of matrix polymers in the growing cell-walls of angiosperms. Annu Rev Plant Physiol Plant Mol Biol 37:165–186

Hall DO (1979) Solar energy use through biology—past, present and future. Sol Energy 22:307–328

Hayashi T, Maclachlan G (1984) Pea xyloglucan and cellulose. I. Macromolecular organization. PlantPhys iol7 5:596–604

He L, Terashima N (1989) Formation and structure of lignin in monocotyledons. I. Selective labeling of the structural units of lignin in rice plant (*Oryza sativa*) with ^{3}H and visualization of their distribution in the tissue by microautoradiography. Mokuzai Gakkaishi 35:116–122

Hector RE, Qureshi N, Hughes SR, Cotta MA (2008) Expression of a heterologous xylose transporter in a *Saccharomyces cerevisiae* strain engineered to utilize xylose improves aerobic xylose consumption. Appl Microbiol Biotechnol 80:675–684

Hemschemeier A, Melis A, Happe T (2009) Analytical approaches to photobiological hydrogen production in unicellular green algae. Photosynth Res 102:523–540

Higuchi T (1985) Biosynthesis of lignin. In: Higuchi T (ed) Biosynthesis and biodegradation of wood components. Orlando Academic Press, Orlando, pp 141–160

Hill J, Nelson E, Tilman D, Polasky S, Tiffany D (2006) Environmental, economic, and energetic costs and benefits of biodiesel and ethanol biofuels. Proc Natl Acad Sci U S A 103: 11206–11210

Ho NWY (1998) Genetically engineered Saccharomyces yeast capable of effective co-fermentation of glucose and xylose. Appl Environ Microbiol 64:1852–1859

Howard R (2003) Lignocellulose biotechnology: issues of bioconversion and enzyme production. AfrJ B iotechnol2: 602–619

Hsieh Y (2009) Alkaline pre-treatment of rice hulls for hydrothermal production of acetic acid. ChemEng R esD es87: 13–18

Hsu TA (1996) Pretreatment of biomass. In: Wyman CE (ed) Handbook on bioethanol, production and utilization. Taylor & Francis, Washington, pp 179–212

Hu Q, Sommerfeld M, Jarvis E, Ghirardi M, Posewitz M, Seibert M, Darzins A (2008) Microalgal triacylglycerols as feedstocks fir biofuel production: perspectives and advances. Plant J 54:621–639

Hwang SS, Lee SJ, Kim HK, Ka JO, Kim KJ, Song HG (2008) Biodegradation and saccharification of wood chips of *Pinus strobus* and *Liriodendron tulipifera* by white rot fungi. J Microbiol Biotechnol 18:1819–1825

JarvisM (2003)C elluloses tacksup.N ature426: 611–612

Jensen JK, Kim H, Cocuron J-C, Orler R, Ralph J, Wilkerson CG (2011) The DUF579 domain containing proteins IRX15 and IRX15-L affect xylan synthesis in Arabidopsis. Plant J 66:387–400

Jin SY, Chen HZ (2007) Near-infrared analysis of the chemical composition of rice straw. Ind Crops Prod 26:207–211

Kaar WE, Holtzapple MT (2000) Using lime pretreatment to facilitate the enzymatic hydrolysis of corn stover. Biomass Bioenerg 18:189–199

Kabel MA (2007) Effect of pretreatment severity on xylan solubility and enzymatic breakdown of the remaining cellulose from wheat straw. Bioresour Technol 98:2034–2042

Karhumaa K, Hahn-Hägerdal B, Gorwa-Grauslund MF (2005) Investigation of limiting metabolic steps in the utilization of xylose by recombinant *Saccharomyces cerevisiae* using metabolic engineering. Yeast 22:359–368

Kataeva IA, Yang S-J, Dam P, Poole II PF, Yin Y, Zhou F, Chou W-C, Xu Y, Goodwin L, Sims DR, Detter JC, Hauser LJ, Westpheling J, Adams MWW (2009) Genome sequence of the anaerobic, thermophilic and cellulolytic bacterium "Anaerocellum thermophilum" DSM 6725. J Bacteriol 191:3760–3761

Keegstra K, Talmadge KW, Bauer WD, Albersheim P (1973) The structure of plant cell walls III. A model of the wall of suspension-cultured sycamore cells based on interconnections of the macromolecular components. Plant Physiol5 1:188–197

Kennedy CJ, Cameron GJ, Sturcova A, Apperley DC, Altaner C, Wess TJ, Jarvis MC (2007) Microfibril diameter in celery collenchyma cellulose: X-ray scattering and NMR evidence. Cellulose14:235 –246

Kim SB, Yum DM, Park SC (2000) Step-change variation in acid concentration in a percolation reactor for hydrolysis of hardwood hemicellulose. Bioresour Technol 72:289–294

Kim TH, Kim JS, Sunwoo C, Lee YY (2003) Pretreatment of corn stover by aqueous ammonia. Bioresour Technol90: 39–47

Kim SR, Ha S-J, Wei N, Oh EJ, Jin Y-S (2012) Simultaneous co-fermentation of mixed sugars: a promising strategy for producing cellulosic ethanol. Trend Biotechnol. doi:10,1016/j.tibtech.2012.01.005

Kishimoto T, Chiba W, Saito K, Fukushima K, Uraki Y, Ubukata M (2010) Influence of syringyl to guaiacyl ratio on the structure of natural and synthetic lignins. J Agric Food Chem 58:895–901

Klemm D, Heublein B, Fink H-P, Bohn A (2005) Cellulose: fascinating biopolymer and sustainable raw material. ChemInform36.doi :10.1002/chin.200536238

Kraakman L, Lemaire K, Ma P, Teunissen AW, Donaton MC, Van Dijck P, Winderickx J, de Winde JH, Thevelein JM (1999) A *Saccharomyces cerevisiae* G-protein coupled receptor, Gpr1, is specifically required for glucose activation of the cAMP pathway during the transition to growth on glucose. Mol Microbiol 32:1002–1012

Kuhar S, Nair LM, Kuhad RC (2008) Pretreatment of lignocellulosic material with fungi capable of higher lignin degradation and lower carbohydrate degradation improves substrate acid hydrolysis and the eventual conversion to ethanol. Can J Microbiol 54:305–313

Kurakake M, Ide N, Komaki T (2007) Biological pretreatment with two bacterial strains for enzymatic hydrolysis of office paper. Curr Microbiol 54:424–428

Larkum AW, Ross IL, Kruse O, Hankamer B (2011) Selection, breeding and engineering of microalgae for bioenergy and biofuel production. Trends Biotechnol 30:198–205

Lee JW, Gwak KS, Park JY, Park MJ, Choi DH, Kwon M, Choi IG (2007) Biological pretreatment of softwood *Pinusde nsifbr a* by three white rot fungi. J Microbiol 45:485–491

Li L, Cheng XF, Leshkevich J, Umezawa T, Harding SA, Chiang VL (2001) The last step of syringyl monolignol biosynthesis in angiosperms is regulated by a novel gene encoding sinapyl alcohol dehydrogenase. Plant Cell 13:1567–1585

Liu E, Hu Y (2011) Construction of a xylose-fermenting *Saccharomyces cerevisiae* strain by combined approaches of genetic engineering, chemical mutagenesis and evolutionary adaptation. Biochem EngJ 48: 204–210

Lynd L, Cushman J, Nichols R, Wyman C (1991) Fuel ethanol from cellulosic biomass. Science 251:1318–1323

Lynd LR, Wyman CE, Gerngross TU (1999) Biocommodity engineering. Biotechnol Prog 15:777–793

Lynd LR, Weimer PJ, van Zyl WH, Pretorius IS (2002) Microbial cellulose utilization: fundamentals and biotechnology. Microbiol Mol Biol Rev 66:506–577

Mabee WE, Gregg DJ, Arato C, Berlin A, Bura R, Gilkes N, Mirochnik O, Pan X, Pye EK, Saddler JN (2006) Updates on softwood-to-ethanol process development. Appl Biochem Biotechnol129–132: 55–70

Mackie KL, Brownell HH, West KL (1985) Effect of sulfur dioxide and sulfuric acid on steam explosion of aspen wood. J Wood Chem Technol 5:405–425

Majid J, Luxhoi J, Lyshede OB (2004) Decomposition of plant residues at low temperatures separates turnover of nitrogen and energy rich tissue components in time. Plant Soil 258:351–365

Manthey FA, Hareland GA, Huseby DJ (1999) Soluble and insoluble dietary fiber content and composition in oat. Cereal Chem76: 417–420

McCann MC, Roberts K (1991) Architecture of the primary cell wall. In: Lloyd CW (ed) The cytoskeletal basis of plant growth and form. Academic Press, Toronto, pp 109–129

McCann MC, Wells B, Roberts K (1990) Direct visualization of cross-links in the primary plant cell wall. J Cell Sci 96:323–334

McCann MC, Wells B, Roberts K (1992) Complexity in the spatial localization and length distribution of plant cell-wall matrix polysaccharides. J Microsc 166:123–136

McMillan JD (1994) Pretreatment of lignocellulosic biomass. In: Himmel ME, Baker JO, Overend RP (eds) Enzymatic conversion of biomass for fuels production. American Chemical Society, Washington, pp 292–324

Melis A, Zhang L, Forestier M, Ghirardi ML, Seibert M (2000) Sustained photobiological hydrogen gas production upon reversible inactivation of oxygen evolution in the green alga *Chlamydomonas reinhardtii*.P lantPhys iol1 22:127–135

Mellerowicz EJ, Baucher M, Sundberg B, Boerjan W (2001) Unraveling cell wall formation in the woody dicot stem. Plant Mol Biol 47:239–274

Michalowicz G, Toussaint B, Vignon MR (1991) Ultrastructural-changing in poplar cell wall during steam explosion treatment. Holzforschung45: 175–179

Miller SS, Fulcher RG (1995) Oat endosperm cell walls: II. hot-water solubilization and enzymatic digestion of the wall. Cereal Chem72: 428–432

Miller S, Hester R (2007) Concentrated acid conversion of pine softwood to sugars. Part 1: use of a twin-screw reactor for hydrolysis pretreatment. Chem Eng Commun 194:85–102

MohnenD (2008)Pe ctins tructurea ndbi osynthesis.C urrO pinP lantB iol11: 266–277

Mosier N, Wyman C, Dale B, Elander R, Lee YY, Holtzapple M, Ladisch M (2005) Features of promising technologies for pretreatment of lignocellulosic biomass. Bioresour Technol 96:673–686

Mueller SC, Brown RM Jr (1980) Evidence for an intramembranous component associated with a cellulose microfibril synthesizing complex in higher plants. J Cell Biol 84:315–326

Muller PR (1974) Look back without anger: a reappraisal of William A. Dunning. J Am Hist 61:325–338

Nakamura N, Yamada R, Katahira S, Tanaka T, Fukuda H, Kondo A (2008) Effective xylose/cellobiose co-fermentation and ethanol production of xylose-assimilating *S. cerevisiae* via expression of β-glucosidaseoni tsc ells urface.E nzymM icrob Technol43: 233–236

Narayanaswamy N, Faik A, Goetz DJ, Gu T (2011) Supercritical carbon dioxide pretreatment of corn stover and switchgrass for lignocellulosic ethanol production. Bioresour Technol 102:6995–7000

Ni H, Laplaza JM, Jeffries TW (2007) Transposon mutagenesis to improve the growth of recombinant *Saccharomycesc erevisiae* on xylose. Appl Environ Microbiol 73:2061–2066

Ohta K, Beall DS, Mejia JP, Shanmugam KT, Ingram LO (1991) Genetic improvement of *Escherichia coli* for ethanol production: chromosomal integration of *Zymomonas mobilis* genes encoding pyruvate decarboxylase and alcohol dehydrogenase II. Appl Environ Microbiol57:893–900

Pan X, Xie D, Kang KY, Yoon SL, Saddler JN (2007) Effect of organosolv ethanol pretreatment variables on physical characteristics of hybrid poplar substrates. Appl Biochem Biotechnol 136–140:367–377

Peng B, Shen Y, Li X, Chen X, Hou J, Bao X (2012) Improvement of xylose fermentation in respiratory-deficient xylose-fermenting *Saccharomycesc erevisiae*.M etabE ng14: 9–18

Perrin RM, Derocher AE, Bar-Peled M, Zeng W, Norambuena L, Orellana A, Raikhel NV, Keegstra K (1999) Xyloglucan fucosyltransferase, an enzyme involved in plant cell wall biosynthesis. Science 284:1976–1979

Persson S, Caffall KH, Freshour G, Hilley MT, Bauer S, Poindexter P, Hahn MG, Mohnen D, Somerville C (2007) The Arabidopsis irregular xylem8 Mutant is deficient in glucuronoxylan and homogalacturonan, which are essential for secondary cell wall integrity. Plant Cell 19:237–255

Persson T, Ren JL, Joelsson E, Jönsson AS (2009) Fractionation of wheat and barley straw to access high molecular-mass hemicelluloses prior to ethanol production. Bioresour Technol 100:3906–3913

Posten C (2009) Design principles of photo-bioreactors for cultivation of microalgae. Eng Life Sci9:165–177

Ralph J, Lundquist K, Brunow G, Lu F, Kim H, Schatz PF, Marita JM, Hatfield RD, Ralph SA, Christensen JH, Boerjan W (2004) Lignins: natural polymers from oxidative coupling of 4-hydroxyphenylpropanoids.P hytochemistry3 :29–60

Ralph J, Brunow G, Harris PJ, Dixon RA, Schatz PF, Boerjan W (2008) Lignification: are lignins biosynthesized via simple combinatorial chemistry or via proteinaceous control and template replication? In: Daayf F, Hadrami A El, Adam L, Ballance GM (eds), Recent advances in polyphenol research, Vol 1. Wiley-Blackwell Publishing, Oxford, pp 36–66

Ranocha P, Chabannes M, Chamayou S, Danoun S, Jauneau A, Boudet A-M, Deborah G (2002) Laccase down-regulation causes alterations in phenolic metabolism and cell wall structure in poplar. Plant Physiol 129:145–155

RegalbutoJ R(2009)C ellulosicbi ofuels—gotg asoline?S cience3 25:822–824

Reiter W-D, Chapple CCS, Somerville CR (1993) Altered growth and cell walls in a fucose-deficient mutant of Arabidopsis. Science 261:1032–1035

Rogalinski T, Ingram T, Brunner GJ (2008) Hydrolysis of lignocellulosic biomass in water under elevated temperatures and pressures. J Supercrit Fluids 47:54–63

Rosgaard L, Pedersen S, Meyer AS (2007) Comparison of different pretreatment strategies for enzymatic hydrolysis of wheat and barley straw. Appl Biochem Biotechnol 143:284–296

RubinEM (2008)G enomicsof c ellulosicbi ofuels.N ature454: 841–845

Saha BC, Cotta MA (2010) Comparison of pretreatment strategies for enzymatic saccharification of barley straw to ethanol. New Biotechnol 27:10–16

Saulnier L, Marot C, Chanliaud E, Thibault JF (1995) Cell wall polysaccharide interactions in maize bran. CarbohydrPol ym26: 279–287

SchellerH V,U lvskovP(2010)H emicelluloses. AnnuR evP lantB iol61: 263–289

ServiceR F(2011) Algae'ss econdt ry.S cience3 33:1238–1239

Sheehan J, Camobreco V, Duffield J, Graboski M, Shapouri H (1998) Life cycle inventory of biodiesel and petroleum diesel for use in an urban bus. National Renewable Energy Lab., Golden,N RELPubl .N o.SR -580-24089

Sheehan J, Aden A, Paustian K, Killian K, Brenner J, Walsh M, Nelson R (2003) Energy and environmental aspects of using corn stover for fuel ethanol. J Ind Ecol **7**:117–146. http://mitpress.mit.edu/jie

Sjostrom E (1993) Wood chemistry: fundamentals and applications, 2nd edn. Academic press, San Diego, p 292

Skinner KA, Leathers TD (2004) Bacterial contaminants of fuel ethanol production. J Ind Microbiol Biotechnol 31:401–408

SomervilleC (2006) Thebi llion-tonbi ofuelsvi sion.S cience312: 1277

Stephenson PG, Moore CM, Terry MJ, Zubkov MV, Bibby TS (2011) Improving photosynthesis for algal biofuels: toward a green revolution. Trends Biotechnol 29:615–623

Sterjiades R, Dean JFD, Eriksson K-EL (1992) Laccase from sycamore maple (*Acev pseudoplatnnus*) polymerizes monolignols.P lantPhys iol9 9:1162–1168

Sterling JD, Atmodjo MA, Inwood SE, Kolli VSK, Quigley HF, Hahn MG, Mohnen D (2006) Functional identification of an Arabidopsis pectin biosynthetic homogalacturonan galactosyltransferase.Proc N atl AcadS ciU S A103: 5236–5241

Sun Y, Cheng JJ (2002) Hydrolysis of lignocellulosic materials for ethanol production: a review. Bioresour Technol83: 1–11

Svetlichnyi VA, Svetlichnaya TP, Chernykh NA, Zavarzin GA (1990) *Anaerocellum thermophilum* gen. nov., sp. nov., an extremely thermophilic cellulolytic eubacterium isolated from hot-springs in the valley of Geysers. Microbiology 59:598–604

Taherzadeh MJ, Karimi K (2008) Pretreatment of lignocellulosic wastes to improve ethanol and biogas production: a review. Int J Mol Sci 9:1621–1651

Thomsen MH, Thygesen A, Thomsen AB (2008) Hydrothermal treatment of wheat straw at pilot plant scale using a three-step reactor system aiming at high hemicellulose recovery, high cellulose digestibility and low lignin hydrolysis. Bioresour Technol99: 4221–4228

Tilman D, Hill J, Lehman C (2006) Carbon-negative biofuels from low-input high-diversity grassland biomass. Science 314:1598–1600

Touzel J-P, Chabbert B, Monties B, Debeire P, Cathala B (2003) Synthesis and characterisation of dehydrogenation polymers in *Gluconacetobacter xylinus* cellulose and cellulose/pectin composite.J AgriF oodC hem51: 981–986

Trumbly RJ (1992) Glucose repression in the yeast *Saccharomyces cerevisiae*. Mol Microbiol 6:15–21

Tsekos I (1999) The sites of cellulose synthesis in algae: diversity and evolution of cellulose-synthesizing enzyme complexes. J Phycol 35:635–655

U. S. DOE (2006) Breaking the biological barriers to cellulosic ethanol: a joint research agenda, DOE/SC/EE-0095, U.S. Department of Energy Office of Science and Office of Energy Efficiency and Renewable Energy

van Maris AJ, Abbott DA, Bellissimi E, van den Brink J, Kuyper M, Luttik MA, Wisselink HW, Scheffers WA, van Dijken JP, Pronk JT (2006) Alcohol fermentation of carbon sources in biomass hydrolysates by *Saccharomyces cerevisiae*, current status. Anttonie Van Leeuwenhoek90 :391–418

Versele M, Thevelein JM, Van Dijck P (2004) The high general stress resistance of the *Saccharomyces cerevisiae* fill adenylate cyclase mutant (CyrI(Lys1682)) is only partially dependent on trehalose, Hsp104 and overexpression of Msn2/4-regulated genes. Yeast 21:75–86

Walfridsson M, Bao X, Anderlund M, Lilius G, Bülow L, Hahn-Hägerdal B (1996) Ethanolic fermentation of xylose with *Saccharomyces cerevisiae* harboring the *Thermus thermophilis xylA* gene, which expresses an active xylose (glucose) isomerase. Appl Environ Microbiol 62:4648–4651

Watanabe T, Koshijima T (1988) Evidence for an ester linkage between lignin and glucuronic-acid in lignin carbohydrate complexes by DDQ-oxidation. Agric Biol Chem 52:2953–2955

Whitney SEC, Brigham JE, Darke A, Reid JSG, Gidley MJ (1995) In vitro assembly of cellulose/xyloglucan networks: ultrastructural and molecular aspects. Plant J 8:491–504

Whitney SEC, Brigham JE, Darke A, Reid JSG, Gidley MJ (1998) Structural aspects of the interaction of mannan-based polysaccharides with bacterial cellulose. Carbohydr Res 307:299–309

Whitney SEC, Gothard MGE, Mitchell JT, Gidley MJ (1999) Roles of cellulose and xyloglucan in determining the mechanical properties of plant cell walls. Plant Physiol 121:657–663

Whitney SEC, Wilson E, Webster J, Bacic A, Gidley MJ (2006) Effects of structural variation in xyloglucan polymers on interactions with bacterial cellulose. Am J Bot 93:1402–1414

Wightman R, Turner S (2010) Trafficking of the plant cellulose synthase complex. Plant Physiol 153:427–432

Wu AM, Rihouey C, Seveno M, Hornblad E, Singh SK, Matsunaga T, Ishii T, Lerouge P, Marchant A (2009) The Arabidopsis IRX10 and IRX10-like glycosyltransferases are critical forgluc uronoxylan biosynthesis during secondary cell wall formation. Plant J 57:718–731

Wu Y, Williams M, Bernard S, Driouich A, Showalter AM, Faik A (2010a) Functional identification of two nonredundant Arabidopsis α(1,2)fucosyltransferases specific to arabinogalactan proteins. J Biol Chem285: 13638–13645

Wu AM, Hornblad E, Voxeur A, Gerber L, Rihouey C, Lerouge P, Marchant A (2010b) Analysis of the Arabidopsis IRX9/IRX9-L and IRX14/IRX14-L pairs of glycosyltransferase genes reveals critical contributions to biosynthesis of the hemicellulose glucuronoxylan. Plant Physiol1 53:542–554

Xu P, Donaldson LA, Gergely ZR, Staehelin LA (2007) Dual-axis electron tomography: a new approach for investigating the spatial organization of wood cellulose microfibrils. Wood Sci Technol41: 101–116

Yang S-J, Kataeva I, Hamilton-Brehm SD, Engle NL, Tschaplinski TJ, Doeppke C, Davis M, Westpheling J, Adams MWW (2009) Efficient degradation of lignocellulosic plant biomass, without pretreatment, by the Thermophilic Anaerobe "Anaerocellum thermophilum" DSM 6725. ApplEn virM icrobiol75: 4762–4769

Yuan JS, Tiller KH, Al-Ahmad H, Stewart NR, Stewart NC (2008) Plants to power: bioenergy to fuel the future. Trends Plant Sci 13:421–429

Zandleven J, Beldman G, Bosveld M, Schols HA, Voragen AGJ (2006) Enzymatic degradation studies of xylogalacturonans from apple and potato, using xylogalacturonan hydrolase. CarbohydrPolym 6 5:495–503

Zhang YHP, Ding ST, Mielenz JR, Cui JB, Elander RT, Laser M, Himmel ME, McMillan JR, Lynd LR (2007) Fractionating recalcitrant lignocellulose at modest reaction conditions. BiotechnolB ioeng97: 214–223

Chapter 2
Thermo-Mechanical Pretreatment of Feedstocks

Chinnadurai Karunanithy and Kasiviswanathan Muthukumarappan

Abstract Pretreatment is the first step to open up the complex structure of feedstock and to facilitate the access of hydrolytic enzymes to carbohydrates. Researchers are paying utmost care to pretreatment since it has pervasive impact in downstream processing. Numerous pretreatments have been investigated in the past with various degrees of success and still several new investigations are under progress. Extrusion/thermo-mechanical process is a continuous viable pretreatment and easy to adapt at industry scale with flexibility in process modifications. Agricultural residues and dedicated energy crops such as corn stover, prairie cord grass, wheat straw, rice straw, rape straw, barley straw, wheat bran, soybean hull, switchgrass, miscanthus, big bluestem, and forest products such as poplar, pine, Douglas fir, and eucalyptus were subjected to extrusion pretreatment with or without chemicals. This chapter provides an overview of extrusion pretreatments including the mechanism influencing extruder and feedstock parameters, evaluation of pretreatment efficiency including production of potential fermentation inhibitors, torque requirements, and direction for future work.

Keywords Extrusion • Screw speed • Barrel temperature • Particle size • Moisture content • Sugar recovery • Hydrolysis • Torque • Fermentation inhibitors

C.K arunanithy(✉)
Fooda ndN utrition,U niversityof Wisconsin-Stout,M enomonie, WI54751 ,U SA
e-mail:c hinnaduraik@uwstout.edu

K.M uthukumarappan
Departmentof Agriculturala ndB iosystemsEngi neering,So uthD akotaS tateU niversity,
Brookings,SD 57007 ,U SA
e-mail:ka s.muthukum@sdstate.edu

T. Gu(e d.), *Green Biomass Pretreatment for Biofuels Production*,
SpringerBriefs in Green Chemistry for Sustainability,
DOI:10.1007/978-94- 007-6052-3_2,© The Author(s)2013

2.1 Introduction

About 60 % of the world's ethanol is produced in the US and Brazil exploiting sugarcane and corn, respectively. Economics and limitation in grain supply lead to search for alternative resources. In order to meet the ever-growing fuel demand, worldwide researchers are exploring alternate renewable resources for more than three decades. Biomass is an attractive feedstock due to its renewable nature, positive environmental impact, and abundant supply. According to the US National Research Council (National Research Council 2000), biobased industry should provide at least 10 and 50 % of liquid fuel by the year 2020 and 2050, respectively. Further, the 2007 Energy Act mandates the production of 21 billion gallons of biofuels from noncorn starch materials by 2022.

Ethanol production from biomass is quite different from corn, because the accessibility of carbohydrates in biomass to hydrolytic enzymes is more difficult than the starch in grain (Gibbons et al. 1986). The complex structure and recalcitrant nature of biomass necessitates an additional step called pretreatment. The primary purposes of pretreatment are to open up the biomass structure, to increase accessible surface area, to reduce the cellulose crystallinity, and to increase the porosity, pore size, and pore volume. In other words, pretreatment should address the factors influencing enzymatic hydrolysis. Yang and Wyman (2008) reported that pretreatment is a key step in converting biomass into biofuel because of its pervasive impact on the downstream processing. Numerous pretreatment methods such as physical, chemical, biological, and their combinations can be found in the literature. Most of these pretreatments often involve high temperature and acidic conditions resulting in degradation and leading to the formation of potential fermentation inhibitors. Moreover, most of the conventional pretreatment methods are batch in nature with low throughput and low solids loading rate resulting in high energy consumption. Dilute acid pretreatment requires corrosion resistant construction materials, neutralization of the pretreated feedstocks; hydrothermal pretreatment requires elevated temperature; chemicals are expensive and difficult to recover and reuse; and in addition safety becomes an important concern. An effective pretreatment must expose the fiber, thereby the cellulose fraction could be hydrolyzed by cellulases without significant trouble, preserve the hemicellulose, and avoid the inhibitory by-products formation (Laser et al. 2002). Inexpensive chemicals, simple equipment and procedures enable the pretreatment to be economical. Thermomechanical/extrusion pretreatment meets the above listed criteria and address the listed issues associated with other pretreatment methods. This chapter presents in-depth a review and discussion on extrusion pretreatment of different feedstocks.

2.2 ExtrusionPr etreatment

2.2.1 Extruder'sC apability

Extrusion is a continuous cost-effective, fast, and simple process hence practical and useful for large-scale operation with high throughput and adaptability for many different process modifications (high pressure, explosion, chemical addition, and reactive extrusion). Extruder screw and barrel can be made using acid-resistant stainless steel and its alloy, such as SS 316L (Chen et al. 2011) and AL6XN (Miller and Hester 2007) even if one has to add acid during extrusion pretreatments. Extrusion pretreatments of different feedstocks revealed that no effluent disposal cost is involved due to the absence of any effluent production, no solid loss, and no safety issues are involved. The unique advantages of extrusion pretreatment are high shear, rapid mixing, short residence time, excellent temperature and screw speed control, and flexibility in screw configuration to mention a few. Moderate temperature employed in extrusion pretreatment prevents the degradation and formation of potential fermentation inhibitors (de Vrije et al. 2002). Extruder can accommodate wide range of feedstock sizes especially larger sizes when compared to other leading pretreatment methods; thereby extruder leads to remarkable saving in terms of size reduction.

2.2.2 Mechanism of Extrusion Pretreatment

In general, as the feedstock pass through an extruder barrel, high shear is exerted by the screw, high pressure and temperatures are developed (Lamsal et al. 2010) causing defibration, fibrillation, and shortening the fibers (change in aspect ratio) (de Vrije et al. 2002). When the pretreated feedstock exits through die section, some of the moisture present in the feedstock is flashed off into steam due to sudden drop in the pressure resulting in expansion and porous structure (de Vrije et al. 2002). Several researchers have attributed possible reasons for the increase in sugar recovery in extrusion pretreatment includes increases in surface area (Karunanithy and Muthukumarappan 2011a, b, c, d), specific surface area (Chen et al. 2011; Lee et al. 2009, 2010), pore size (Jurisic et al. 2009; Zhang et al. 2012a), and pore quantity (Zhang et al. 2012b), a decrease in cellulose crystallinity (Lamsal et al. 2010; Zhang et al. 2012b), and induction of micro/nano fiballization (Zhang et al. 2012b), all of these facilitate the access of enzymes to cellulose. The developed mechanical high shear breaks down the structure of feedstock and facilitates contact between feedstock and chemicals (acid/alkali/cellulose affinity additives) addcd during extrusion due to effective mixing. Further, the shear exerted by the screw helps in continuous removal of softened regions of the feedstock and expose fresh interior surface to chemical and thermal action that improves the overall rate of decrystallization (Lamsal et al. 2010).

2.3 Extruder Parameters Influencing Sugar Recovery

Pretreatment efficiency (sugar recovery) depends on several extruder parameters such as compression ratio, screw configuration, screw speed, and barrel temperature and they are presented below.

2.3.1 ScrewC ompressionR atio

Compression ratio is the channel depth at feed to channel depth at discharge that has a direct impact on shear development within the extruder barrel; it affects sugar release from the feedstock due to the process of plasticization which occurs in the compression zone. Harper (1981) listed several ways to achieve the desired compression ratio by varying the screw and barrel configuration: by increasing the root diameter, by decreasing the pitch or barrel diameter with constant root diameter, by decreasing the screw pitch in a decreasing barrel diameter, and by introducing restrictions. However, the most common ways to achieve compression are gradual decrease of flight depth in the direction of the discharge and a decrease in the pitch in the compression section (Harper 1981). Among the literature surveyed, only researchers from South Dakota State University studied the effect of screw compression ratio on different feedstocks (Karunanithy and Muthukumarappan 2010a, b). Though screw compression ratio used in food industries typically ranges from 1:1 to 5:1 that are widely available, the authors used only 2:1 and 3:1 in their study to understand the influence of screw compression ratio on sugar recovery. As Harper (1981) described,

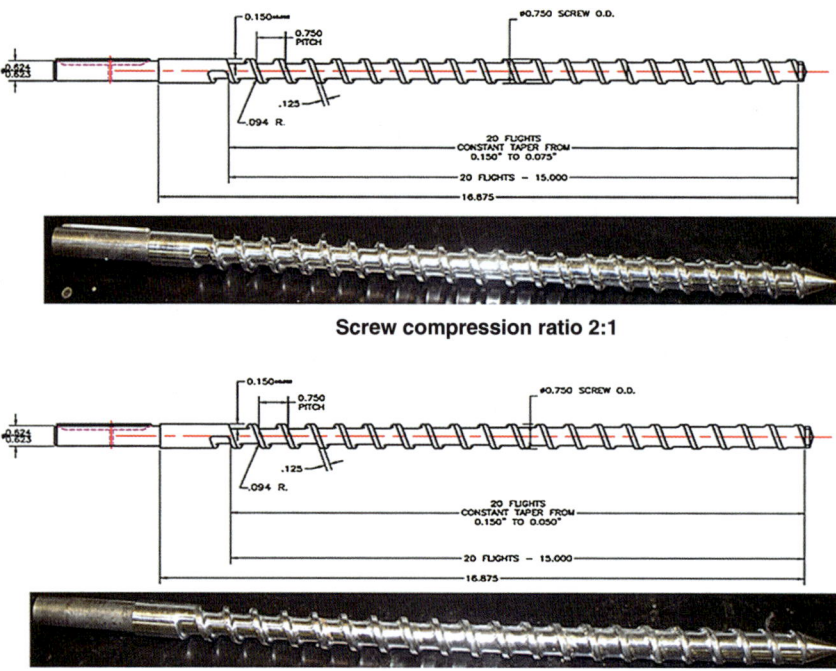

Screw compression ratio 2:1

Screw compression ratio 3:1

Fig. 2.1 Screw diagram showing the difference between 3:1 and 2:1 compression ratio (with permission: C.W. Brabender Instruments Inc, NJ). Reprinted from BioEnergy research, a comparative study on torque requirement during extrusion pretreatment of different feedstocks, © 2011, Chinnadurai Karunanithy, 5, with kind permission from Springer Science+Business Media and any original (first) copyright notice displayed with material

the screws were identical except for the depth of the discharge flight at the screw end and had a continuous taper (decrease of flight depth) from the feed to the discharge of 3.81–1.27 mm and 3.81–1.90 mm for the screws with compression ratios of 3:1 and 2:1, respectively (Fig. 2.1). They concluded that a screw compression ratio of 3:1 was better than 2:1 due to more work (shear force and long residence time) applied on the feedstocks. Recently, Chen et al. (2011) used twin screw extruder with a screw compression ratio of 5:1; however, no comparison was possible due to lack of data with other screw compression ratios.

2.3.2 ScrewC onfiguration

It is an important factor that would affect pretreatment efficiency. In order to use different screw configurations in a single screw extruder, one has to change different screw since the screw shaft and flights are made of single solid rod/shaft or inseparable. In the case of twin screw extruder, one has the flexibility of arranging different screw elements such as forward conveying, forward kneading, eccentric kneading, neutral kneading, mono or bilobal kneading, and reverse kneading in any order for maximizing pretreatment efficiency or other objectives. Different screw elements and screws made thereof is shown in Fig. 2.2. In general, screw elements are arranged to have a constantly decreasing pitch to enhance the shear on the feedstock (Lamsal et al. 2010). N'Diaye et al. (1996) used twin screw extruder for hemicellulose extraction from poplar and they arranged the conveying screw elements to facilitate conveying and heating due to decrease in pitch for compacting, removing of air, and starting impregnation of alkali; followed by bilobal and eccentric kneading elements for homogenizing the slurry and increasing impregnation of alkali; then reverse screw element for separating liquid through the filter and finally conveying screw elements. de Vrije et al. (2002) used twin screw extruder with conveying and reverse screw elements for accumulating and compressing miscanthus between elements space that resulted in high shear development. Lee et al. (2009) arranged left and right corrugated conveying elements for continuous movement of Douglas fir followed by reverse screw elements for increasing the residence time as well pulverization effect and finally conveying elements for forward conveying. One such arrangement is shown in Fig. 2.3 used for soybean hull pretreatment.

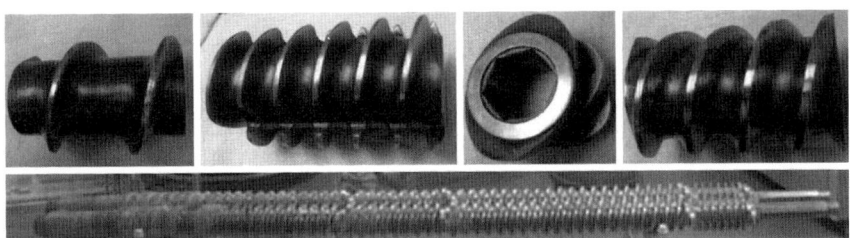

Fig.2.2 Different screw elements and screws made thereof

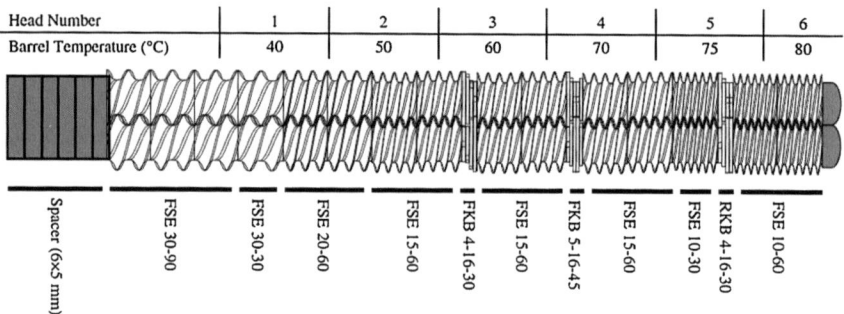

FSE: forward conveying screw element (all double flighted, intermeshing)
FKB: forward kneading block; RKB: reverse kneading block
Numbers on screw elements: pitch (mm)-element length (mm)
Numbers on kneading blocks: number of disks-total block length (mm)-staggering angle of disks

Fig. 2.3 Schematic showing lab-scale twin screw extruder screw profile and barrel temperature setting (Yoo et al. 2012). Reprinted from applied biochemistry and biotechnology, Soybean hulls pretreated using thermo-mechanical extrusion—hydrolysis efficiency, fermentation inhibitors, and ethanol yield, © 2011, Juhyun Yoo, 2, with kind permission from Springer Science+Business Media and any original (first) copyright notice displayed with material

Chen et al. (2011) used twin screw extruder with different screw elements such as conveying, kneading, and combing mixer blocks for transport, mixing, and granulation of rice straw, respectively. Recently, Choi and Oh (2012) employed twin screw extruder with forward conveying and kneading screw elements for accomplishing effective conveying, mixing, and pretreatment of rape straw. Researchers from University of Nebraska, Lincoln has used different screw configuration (Fig. 2.4) with conveying element, reverse element for enhancing mixing and pulverizing, and again conveying element for creating more pressure (Zhang et al. 2012a, b).

2.3.3 ScrewSpe ed

The authors reported that *screw* speed is responsible for the rate of shear development and the mean residence time (Karunanithy and Muthukumarappan 2010a, b), wherein these are two sides of a coin and competing factors (Yoo et al. 2012). In general, the screw speed is inversely proportional to the mean residence time of the feedstock in the extruder (Chen et al. 2011; Lamsal et al. 2010). Researchers have reported that the effect of screw speed was inconsistent with respect to sugar recovery from different feedstocks.

Soybean hull pretreated using twin screw extruder exhibited mixed behavior depending upon the barrel temperature i.e., sugar recovery decreased with an increase in screw speed (220–420 rpm) at 110 °C, whereas sugar recovery increased with an increase in screw speed at 150 °C (Lamsal et al. 2010). In another soybean hull twin screw extrusion study, an increase in screw speed from 220 to 350 rpm increased sugar recovery and further increase to 420 rpm

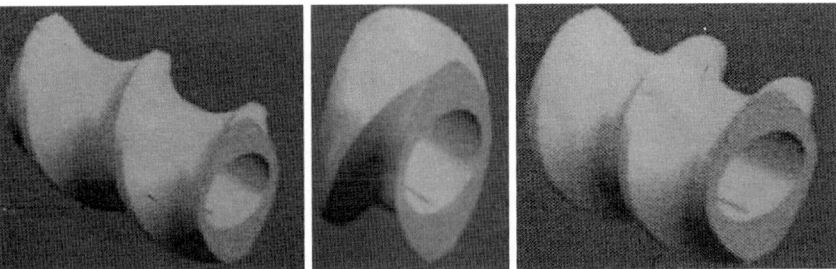

Fig. 2.4 Screw configurations used for corn stover pretreatment (Zhang et al. 2012b). Reprinted from applied biochemistry and biotechnology, pretreatment of corn stover with twin-screw extrusion followed by enzymatic saccharification, © 2011, Shujing Zhang, 2, with kind permission from Springer Science and Business Media and any original (first) copyright notice displayed with material

decreased sugar recovery (Yoo et al. 2011). Similarly, rice straw pretreated in a twin screw extruder with dilute acid also revealed that an increase in screw speed from 30 to 80 rpm increased the sugar recovery and further increase to 150 rpm decreased the sugar recovery (Chen et al. 2011). Twin screw extrusion pretreatment of corn stover revealed that screw speed significantly influenced sugar recovery (Zhang et al. 2012b). In addition, they reported that shear rate developed at 80 rpm would be sufficient for lignin softening, further increase in screw speed lessened the residence time that possibly resulted in insufficient destruction of lignin sheath. The possible reasons are as lower screw speed allows more time for hydration, thermal degradation, and mechanical separation of lignin and cellulosic components at the same time reduces mechanical energy input and less intensive treatment. Since residence time and mechanical energy input are competing factors, their effects might mitigate each other to some extent (Yoo et al. 2011). Therefore, one has to identify whether residence time or rate of shear development plays a major role in the pretreatment of specific feedstocks. Accordingly, Karunanithy and Muthukumarappan (2010a, b); Karunanithy et al. (2012) found that the rate of shear development (high screw speed) was critical for corn stover, switchgrass, and pine chips and the mean residence time was critical for big bluestem and prairic cord grass.

Karunanithy and Muthukumarappan (2010a, b, c, 2011a, b, c, d, e, f, g, h); Karunanithy et al. (2012) observed that screw speed had a strong influence on sugar recovery from variety of feedstocks. Screw speed had a positive influence on sugar recovery from different feedstocks such as corn stover, switchgrass and big bluestem when pretreated in a single screw extruder, whereas it exerted a negative impact on sugar recovery when alkali soaked corn stover, switchgrass, big bluestem, and prairie cord grass pretreated in a single screw extruder. More energy is available for fiber breakage as the screw speed increases that the fiber length and aspect ratio. Moreover, the applied high shear forces during extrusion influences the fiber length reduction, which is similar to pelletizing (Baillif and Oksman 2009). The reduction in fiber length increases the accessible surface area for enzymes during hydrolysis (as presented in Sects. 2.6.1 and 2.6.2). Though the linear term of screw speed did not have significant effect on sugar recovery from prairie cord grass (Karunanithy and Muthukumarappan 2011d) and miscanthus (Jurisic 2012), its quadratic term and interaction with other variables had a significant effect. Screw speed influenced sugar recovery from corn stover when pretreated in twin screw extruder without alkali, whereas it was not an influencing factor when alkali was added for moisture balancing and extruded in a twin screw extruder (Zhang et al. 2012a, b). Possible reasons might be that the differences in screw speeds (40–140 rpm vs. 40–100 rpm) and moisture content (22.5–27.5 vs. 50 % wb). These studies also emphasis that the screw speed is one of the factors influencing the pretreatment efficiency.

2.3.4 Barrel Temperature

Extruder barrel temperature is another important extrusion parameter that facilitates the melting or softening/plasticizing of the feedstock; and thereby affects the flow pattern and residence time. Though there is several extrusion pretreatments literature found, only a few studies the effect of barrel temperature on sugar recovery. Lee et al. (2009) achieved fine and even fibrillated (most of them <1 μm and some of them <100 μm) Douglas fir fibers when extruded in a twin screw extruder at a barrel temperature of 40 °C as compared to 120 °C. They attributed the fibrillation to high torque at low temperature. Researchers from South Dakota State University have investigated the temperature effect with a wide range of temperature 45–225 °C for corn stover, switchgrass, big bluestem, and prairie cord grass using a single screw extruder with or without alkali soaking. The authors observed that the barrel temperature had a positive influence on the sugar recovery from switchgrass and big bluestem when pretreated with or without alkali soaking (Karunanithy and Muthukumarappan 2011a, b, c, d). They also reported that barrel temperature had a positive effect on sugar recovery for corn stover without alkali soaking, whereas it had negative effect when alkali soaked and extruded in a single screw extruder. Although barrel temperature was not an influencing factor for prairie cord grass without alkali soaking, it turned to be an influencing factor when prairie cord grass alkali soaked and extruded. Optimization extrusion pretreatment conditions for different feedstocks revealed that not only linear term of barrel temperature affected sugar recoveries but also its quadratic term and interaction with other factors (Karunanithy and Muthukumarappan 2011a, b, c, d, e, f, g, h). Recently, Jurisic (2012) and Karunanithy et al. (2012) observed that the barrel temperature had a strong impact on sugar recovery from miscanthus and pine chips, respectively. On one hand, an increase in the barrel temperature would enhance the moisture evaporation that develops more friction thereby more disturbances to the feedstocks. On the other hand, an increase in the barrel temperature would decrease viscosity, which resulted in a more flowable material and decrease in residence time. Chen et al. (2011) reported an increase in barrel temperature (150–155 °C) increased sugar recovery and further increase (170 °C) led to decrease in sugar recovery from rape straw pretreated in a twin screw extruder. They (Chen et al. 2011) also reported a similar trend when rice straw pretreated in a twin screw extruder, as the barrel temperature increased from 80 to 120 °C sugar recovery increased and further increase to 160 °C decreased the sugar recover. A possible reason is that degradation of sugar into by-products occurs at high severity factor (high temperature and acid concentration).

2.4 FeedstockP arametersI nfluencing Sugar Recovery

Feedstock parameters such as type of feedstocks, moisture content, and particle size affect not only the sugar recovery but also economics. The following section presents the details of them.

2.4.1 TypeofF eedstock

Optimum pretreatment conditions depends upon the type of feedstocks especially the chemical composition including lignin content. The chemical composition is different for agricultural residues, perennial grasses, and wood species as shown in Table 2.1. Lignin is considered as key hurdle in the conversion process and it varied from 2 to 34.7 % depending upon the feedstocks. In general, the low cellulose and high lignin would affect the pretreatment efficiency (sugar recovery). Literature survey reveals that there was no study conducted to understand the effect of feedstock's extractives on extrusion pretreatment and the impact of ash content on wear and tear of the extruder screw and barrel and they are potential topics that should be addressed before adopting at industrial scale. So far several researchers have optimized extrusion parameters for individual feedstocks, whereas finding the optimum pretreatment conditions for mixed feedstocks is difficult. In reality, a biorefinery plant may get a mixed feedstocks from different farms, wherein separating thcm is a difficult task. This indicates the research opportunitiesforoptimiz ingpre treatmentc onditionsformix edfe edstocks.

2.4.2 FeedstockM oistureC ontent

Moisture content is an important factor in most of the mechanical and physicochemical pretreatments such as disc refining, wet ball milling, steam pretreatment, and steam explosion used for softening the matrix of feedstocks. In extrusion pretreatment, feedstock moisture content plays a role in the development of friction and residence time, in addition to thermal softening by utilizing the barrel

Table 2.1 Chemical composition of different feedstocks used in extrusion pretreatments (% dry basis)

Feedstock	Cellulose	Hemicellulose	Lignin	Ash	Extractives	References
Corn stover	33.2	22.0	14.9	10.9	12.9	Zhang et al. (2012a, b)
Corn stover	35.2–38.4	19.0–21.6	19.2–21.2	6.6–7.2	8.8–12.5	Karunanithy and Muthukumarappan (2012a)
Switchgrass	25.5–31.3	22.3–25.6	24.7–26.8	3.0–5.6	18.5–23.8	Karunanithy and Muthukumarappan (2012a)
Big bluestem	34.0–36.5	17.3–19.3	19.1–21.1	9.2–11.2	12.5–19.2	Karunanithy and Muthukumarappan (2012a)
Prairie cord grass	33.1–33.5	13.1–17.7	20.5–21.5	5.0–5.6	20.3–22.	Karunanithy and Muthukumarappan (2012a)
Miscanthus	38.2	24.3	25.0	2.0	5.6	de Vrije et al. (2002)
Miscanthus	34.3	36.8	24.8	4.9	5.7	Jurisic (2012)
Rice straw	38.8	12.3	19.5	13.1	13.1	Chen et al. (2011)
Barely straw	39.1	25.7	16.4	6.8	10.2	Duque et al. (2011)
Rape straw	32.9	17.1	NR	NR	NR	Choi and Oh (2012)
Wheat bran	10	37	7	NR	NR	Lamsal et al. (2010)
Soybean hull	31	19	14	NR	NR	Lamsal et al. (2010)
Soybean hull	36.2	17.2	2	NR	NR	Yoo et al. (2011)
Poplar	44.9	36.8	17.7	0.19	5.0	N'Diaye et al. (1996)
Douglas fir	46.8	28.9	25.3	NR	NR	Lee et al. (2010)
Eucalyptus	42.1	34.6	28.8	NR	NR	Lee et al. (2010)
Pine	33.8	27.6	34.7	0.5	3.8	Karunanithy et al. (2012)

NR not reported

temperature and rate of shear development. In a single screw extruder, the feed-stock is mainly conveyed by friction (Yeh and Jaw 1998), therefore grooved barrel is required for conveying action (Akdogan 1996). Feedstock with low moisture content enhances friction development due to resistance offered by the feedstock. An increase in moisture content would decrease the friction between the feed-stock, screw shaft, and extruder barrel (Chen et al. 2010) due to softening of the cellulose, hemicellulose, lignin complex of the feedstock; moreover, high mois-ture acts as lubricant (Hayashi et al. 1992) contributes to slippage. Because of the above facts, as expected when corn stover, switchgrass, big bluestem, prairie cord grass, and pine wood chips with a wide range of moisture contents (10–50 % wb) were pretreated in a single screw extruder, the sugar recovery decreased remark-ably (Karunanithy and Muthukumarappan 2011a, b, c, d). Optimization of extru-sion pretreatment conditions for many feedstocks revealed that the moisture content has a strong influence on sugar recovery i.e., the lower the moisture content, better the sugar recovery. Most of the ground feedstocks would have moisture content in the range of 5–8 % wb; however, feedstocks with these low moisture contents cannot be pretreated in an extruder since it would get struck in the extruder barrel. Therefore, addition of moisture becomes inevitable. Instead of adding water to bring the feedstock moisture content to a target level, it is better to use the feedstocks from the field with the desired range of moisture thereby one can save drying cost.

Researchers from University of Nebraska, Lincoln reported that twin screw extrusion of corn stover with a moisture content of 22.5–27.5 % wb did not con-tribute to the sugar recovery due to narrow range of moisture content (Zhang et al. 2012b). However, in another study these authors adjusted corn stover moisture content to 50 % wb by adding required alkali solution (Zhang et al. 2012a). Twin screw extrusion of soybean hull with high in-barrel moisture contents (45–50 % wb) resulted in a better sugar recovery. In most of the twin screw extrusion pre-treatment, chemical solutions are added to different feedstocks during extrusion that brings up the moisture content. When compared to single screw extruder, the feedstock moisture content was high in twin screw extruder. A possible reason might be the difference in conveying mechanisms of single screw and twin screw extruders. Akodagan (1996) reported that the feedstock is transferred in bulk from one screw to the other that facilitates the forward conveyance of the feedstock in a twin screw extruder. Again literature survcy indicated that only a few studies have investigated the effect of moisture content on sugar recovery. Since mois-ture content is a critical factor in extrusion pretreatment, optimum range should be identified for each feedstock and for the type of extruder.

2.4.3 Feedstock Particle Size

Feedstock particle size depends on the type of pretreatment methods e.g. most of the chemical pretreatments require 1–2 mm particle size. According to an US patent 5677154 (Draanen and Mello 1997), for ethanol production process requires feedstock particle size in the range of 1–6 mm. Smaller particle sizes contribute to higher surface to mass ratio that are more readily hydrolyzed by enzymes. Cadoche and Lopez (1989) reported that the power requirement for size reduction of feedstocks depends on the final particle size and characteristics of feedstocks. In general, size reduction costs and energy requirements has inverse relation with particle sizes. Particle size not only affect the cost of size reduction but also affects the diffusion kinetics and effectiveness of pretreatment (Kim and Lee 2002; Chundawat et al. 2007), sugar yield (Chang et al. 2001; Hu et al. 2008; Yang et al. 2008), lignin removal (Hu et al. 2008), hydrolysis rate, rheological properties (Chundawat et al. 2007; Desari and Bersin 2007), and acetic acid formation (Guo et al. 2008). The authors investigated the effect of feedstock particle size (2–10 mm) as part of the optimization study and found an 8 mm particle size can be used in single screw extruder without compromising sugar recovery (Karunanithy and Muthukumarappan 2011a, b, c, d). Jurisic (2012) also evaluated the effect of miscanthus particle size on sugar recovery but in a narrow range (0.67–2.33 mm) using a single screw extruder. The effect of feedstock particle size in a twin screw extruder was not found in the literature; therefore, it is another research area that should be addressed before moving into industrial scale.

2.5 ExtrusionPr etreatmentofD ifferentF eedstocks

Researchers evaluated the pretreatment efficiency in terms of ruminant digestibility, sugar recovery, and enzymatic digestibility. Summary of extrusion pretreatment of different feedstocks is divided before and after the year 2000 and presented in the following section.

2.5.1 EarlierE xtrusionStudie s

In the early days, extrusion has been used as means to enhance delignification, ruminant digestibility, nutritive value, and fermentability. Mostly extrusion was used in combination with chemicals such as dilute sulfuric acid (Noon and Hochstetler 1982), sodium hydroxide, sodium sulfide, anthraquinone, anthrahydroquinone, hexamethylenediamine, hexamethylenetetramine, hydrogen peroxide, and ferrous ammonium sulfate (Carr and Doane 1984), and alkaline hydrogen peroxide (Gould et al. 1989; Helming et al. 1989) for improving nutritive value or digestibility of straw/biomass. Noon and Hoschtetler (1982) utilized pilot-scale single screw extruder with dilute sulfuric acid injection at three different locations along the barrel for exploring alcohol fuel from saw dust, corn stover and wheat straw. The authors fed the moisture balanced feedstocks (40 % wb) to the extruder, acid (1.78 % of gross mass flow) was injected about hallway or more down the barrel and they reported 40, 33, and 33 % cellulose conversion for saw dust, corn stover, and wheat straw, respectively. Extrusion treatments (stainless steel interrupted-flight screw and stationary pins) of wheat straw (20 % solids loading rate, 98 °C, 5.6 ± 1 min; 35 rpm) with sodium hydroxide (15.7 % DM), sodium hydroxide (15.7 %) and anthrahydroquinone (0.3 %), and sodium hydroxide (12.7 %) and sodium sulfide (5.0 %) resulted in a lignin removal of 64–72 % and pentosans of 36–43 %, whereas they were 46–56 and 23–27 %, respectively, for without extrusion. The cellulase treatment of the residues converted 90–92 % of the cellulose to glucose compared to 61–69 % for without extrusion (Carr and Doane 1984). The same authors carried out another experiment to disrupt wheat straw lignin, hemicellulose, cellulose complex in a twin screw compounder (200–450 rpm, 103 °C) with sodium hydroxide (0–24 %), sodium hydroxide and anthraquinone (0–0.5 %) and water (1:1–1:9) only. Twin screw compounder treatment of wheat straw with 16 % sodium hydroxide and 30 s residence time resulted in a lignin removal of 62–64 % and cellulose to glucose conversion of 87–92 % (Carra ndD oane1984).

Earlier studies using alkaline hydrogen peroxide (H_2O_2) on the digestibility of wheat straw and similar materials employed as high as 250 g H_2O_2/kg of straw to be considered nonpractical for a commercial process (Gould 1984, 1985; Kerley et al. 1985, 1986; Lewis et al. 1987). Significant amount of liquid stream was also produced with good amount of solubilized hemicellulose. In order to reduce the amount of chemical used and liquid stream, Gould et al. (1989) used an extruder (with flighted screw sections separated by flight interrupting steam-locks) for treating biomass and they reported the addition of H_2O_2 and sodium hydroxide (NaOH) to wheat straw (1.6 cm) in the extruder. The highest in-situ digestibility of more than 75 % was reported with 25 g H_2O_2/kg of wheat straw coupled with extruder and it was tenfold reduction in the amount of H_2O_2 though the digestibility was only slightly lower than the highest values obtained in earlier studies (Kerley et al. 1985, 1986). Helming et al. (1989) reported synergistic effects between alkaline hydrogen peroxide and twin screw extrusion treatments of wheat straw. Chopped wheat straw (1.5–2.5 cm, 50 % moisture content) impregnated in different amounts of NaOH, H_2O_2, and sodium silicate before extruding in a twin screw extruder. By treating wheat straw with 1–2 % H_2O_2 and 4–5 % NaOH in a twin screw extruder, the amount of cell-wall available for microbiological digestion increased from 30 to 40 % for untreated substrate to over 80 % for the treated.

Xylan-Delignification-Process (XDP) utilizes extrusion technology in conjunction with alkali soak and H_2O_2 injection. Chen and Wayman (1989) also employed this sort of treatment for aspen wood chips adding SO_2 prior to extrusion followed by soaking in a 4 % hot NaOH solution as post-extrusion delignification process. When the pretreated wood chip stream was added to water and hydrolysed using cellulase resulted in 10.7 g/l ethanol during a simultaneous saccharification and fermentation with 31.9 % yield of ethanol from dry matter (Tyson 1993). The xylan method continuously treats the biomass by first reacting with a medium containing an alkali agent (pH 11.5), which softens lignin and allows water to enter the biomass. The high temperature and pressure allow the minimization of chemicals as compared to other technologies for cellulose pretreatment- low acid and high temperature or high acid and low temperature. The extrusion mixes, grinds, sterilize, and disrupts the biomass cell-wall. Dale et al. (1995) employed XDP on switchgrass and corn stalk at the extrusion temperature of 200 °C and pressure of 300–400 psi. The authors first added 50 % NaOH solution to biomass and it was 3 % of the biomass weight. After mixing for 15 min, it was fed to extruder and 1 % w/w H_2O_2 (9 % solution) was injected near the entrance of the extruder. The residence time was about 15 s. Authors added H_2O_2 to the extruder barrel to catalyze the breakdown of the biomass structure. They defined the yield as grams of cellobiose and glucose released per gram of cellulose and it was about 85 % within 7 h of enzymatic hydrolysis. The concentration of 4.8 g/l glucose, 3.2 g/l cellobiose, and 2.6 g/l xylose were noted in 24 h for the treated corn stalks. They used ground corn stover (40 mesh) as control and reported a yield of 50 and 72 % in 7 and 24 h, respectively. The treated switchgrass yielded 65 % in 7 h and it increased to 78 % in 24 h, whereas α- cellulose (control) showed 97 % yield.

Williams et al. (1997) conducted an experiment to find out the effects of extrusion on different feedstocks (corn silage and wheat straw) cell-wall quality and to determine whether the fermentability of their cell-wall can be altered. Wheat straw and corn silage were extruded in conical, corotating twin screw extruder at two temperature conditions (60–115–140–150 and 60–135–165–185 °C) and three screw speeds (10, 25, and 40 rpm) without die. The authors tested the extrudates for their fermentability in cumulative gas production method and compared with each other and controlled too. It was reported that an analysis of the cumulative gas production showed significant differences between substrates. Extrusion treatment also led to significant differences, though the effect of screw speed and temperature were not always consistent.

Dale et al. (1999) explored corotating, self-wiping twin screw extruder for ammonia fiber expansion (AFEX). The sugar yield of extrusion AFEX treated corn stover after enzymatic hydrolysis increased up to 3.5 times than that of control. The ruminant digestibility of corn stover increased up to 32 % (from 54 to 71 %) over completely untreated material, and 23 % (from 63 to 77 %) over material extruded without ammonia.

2.5.2 RecentE xtrusionStudie s

It is very difficult to compare the results of the extrusion studies due to differences in the type of extruder, type of feedstocks, chemicals used, enzyme dose, ratios between cellulase and β-glucosidase, and enzymatic hydrolysis conditions. However, Tables 2.2 and 2.3 is presented here for comprehensive understanding and comparisons of different extrusion studies without and with chemicals. Though, this book primarily deals with 'green pretreatments', in order to provide up to date development on extrusion pretreatments the authors included extrusion pretreatments with chemicals. As observed from Tables 2.2 and 2.3, extrusion

Table 2.2 Comparison of recent extrusion pretreatment of different feedstocks without chemicals

Extruder-type	Conditions	Feedstock condition	Hydrolysis condition	Sugar recoveries	References
Corotating twin screw extruder (no die)	Zone 1 and 2 set at 50 and 140 °C, 40–140 rpm, 0.22 kg/h	Corn stover, <2 mm, 22.5–27.5 % wb	Cellic ctec 2: 0.028 g enzy/g dm, 50 °C, 72 h	Glucose 48.8 %, xylose 25 %, and combined 40 %	Zhang et al. (2012b)
Twin screw extruder, l/d 30:1, 18 mm screw diameter, 2.4 mm circular die	Different heads set at 40–80 °C, 280–420 rpm, 0.54–0.6 kg/h	Soy hull, <1.04 mm, 40–50 % wb	Celluclast 1.5L: Novo 188, 0.2:1; 25 FPU/g cellulose, 50 °C, 72 h, 150 rpm	74 % glucose	Yoo et al. (2012)
Twin screw extruder, l/d 30:1, 18 mm screw diameter, 2.6 mm circular die	Different heads set at 40–150 °C, 220–420 rpm, 1.5 kg/h	Wheat bran, soy hull, 25 % wb	47 mg cellulose/g substrate, 50 °C, 120 rpm, 30 min	41–60 % glucose	Lamsal et al. (2010)
Corotating twin-screw extruder	50 °C, 100 rpm, 1 straw:2 water	Wheat straw	16 h	31 % glucose	Ng et al. (2011)
Single screw extruder l/d 20:1, 3:1 screw compression ratio (no die)	45–225 °C, 20–200 rpm	Big bluestem, 2–10 mm, 10–50 % wb	15 FPU Celluclast 1.5L and 60 CBU Novo 188/g dm, 50 °C, 72 h, 150 rpm	41–72 % glucose, 41–90 % xylose, 37–67 % combined	Karunanithy and Muthukumarappan (2011c)
Single screw extruder l/d 20:1, 3:1 screw compression ratio (no die)	45–225 °C, 20–200 rpm	Switchgrass, 2–10 mm, 10–50 % wb	15 FPU Celluclast 1.5L and 60 CBU Novo 188/g dm, 50 °C, 72 h, 150 rpm	27–40 % glucose, 24–60 % xylose, 29–48 % combined	Karunanithy and Muthukumarappan (2011b)
Single screw extruder l/d 20:1, 3:1 screw compression ratio (no die)	45–225 °C, 20–200 rpm	Corn stover, 2–10 mm, 10–50 % wb	15 FPU Celluclast 1.5L and 60 CBU Novo 188/g dm, 50 °C, 72 h, 150 rpm	68–89 % glucose, 79–90 % xylose, 74–90 % combined	Karunanithy and Muthukumarappan (2011a)

dm dry matter

(continued)

Table 2.2 (continued)

Extruder-type	Conditions	Feedstock condition	Hydrolysis condition	Sugar recoveries	References
Single screw extruder l/d 20:1, 3:1 screw compression ratio (no die)	45–225 °C, 20–200 rpm	Prairie cord grass, 2–10 mm, 10–50 % wb	15 FPU Celluclast 1.5L and 60 CBU Novo 188/g dm, 50 °C, 72 h, 150 rpm	41–55 % glucose, 77–85 % xylose, 50–63 % combined	Karunanithy and Muthukumarappan (2011d)
Single screw extruder l/d 20:1, 3:1 screw compression ratio (no die)	100–180 °C, 100–200 rpm	Pine chips, 8 mm, 25–45 % wb	15 FPU Celluclast 1.5L and 60 CBU Novo 188/g dm, 50 °C, 72 h, 150 rpm	35–66 % glucose, 25–68 % xylose, 32–66 % combined	Karunanithy et al. (2012)
Single screw extruder l/d 20:1, 3:1 screw compression ratio (no die)	83–167 °C, 83–167 rpm	Miscanthus, 0.67–2.33 mm, 13.3–21.6 % wb	5 FPU Celluclast 1.5L and 18 CBU Novo 188/g dm, 50 °C, 72 h, 150 rpm	17–78 % glucose	Jurisic (2012)

Table 2.3 Comparison of recent extrusion pretreatment of different feedstocks with chemicals

Extruder-type	Conditions	Chemical(s)	Feedstock condition	Hydrolysis condition	Sugar recoveries	References
Counter rotating twin screw extruder, l/d 17, barrel diameter 15.4 mm	40–180 °C, 50 rpm, 0.09 kg/h	Ethylene glycol, glycerol, dimethyl sulfoxide	Douglas fir, <0.2 mm	40 mg enzyme/g substrate, 45 °C, 250 rpm, 48 h	62.4 % glucose	Lee et al. (2009)
Twin-screw extruder	50–100 °C, 150 rpm	2.75–7.5 % NaOH/DM	Barley straw		89.9 % glucose and 83.5 % xylose	Duque et al. (2011)
Corotating twin-screw extruder, 1.25 m length and 55 mm screw diameter	100 °C, 100 rpm, 15.30 kg/h	12 % NaOH, 70 °C, 4 h	Miscanhus	1.6 FPU Celluclast 1.5L, 2.3 Novo/ g dm, 45 °C, 72 h	69 % glucose, 38 % xylose	de Vrije et al. (2002)
Twin-screw extruder l/d 27.8, 1000 mm length and 36 mm screw diameter, barrel and screw were made of SS316L	150–170 °C, 19.7 min, 7 min reaction time, 0.080 kg/h	1.5–3.5 % H_2SO_4	Rape straw, 1.4–2.36 mm, solid to liquid ratio 1:13–1:9	30 FPU and 70 pNPG/ g cellulose, 50 °C, 120 rpm, 48 h	70.9 % glucose	Choi and Oh (2012)
Corotating twin screw extruder, l/d 56, diameter 50 mm, compression ratio 5:1	80–160 °C, 30–150 rpm, 50 % slr, 5–7 kg/h	1–3 % w/w H_2SO_4	Rice straw, <1 cm	20 FPU/g cellulose, 50 °C, 100 rpm, 72 h	32.9 % glucose	Chen et al. (2011)
Corotating twin screw extruder, l/d 24	Zone 1 and 2 set at 50 and 140 °C, 40–100 rpm, 0.22 kg/h	1–14 % NaOH w/w added to adjust moisture content to 50 % db	Corn stover, <2 mm, 50 % db	Celic Ctec 2: 0.028 g enzy/g dm, 50 °C, 72 h	86.8 % glucose, 50.5 % xylose	Zhang et al. (2012a)

dm dry matter

(continued)

Table 2.3 (continued)

Extruder-type	Conditions	Chemical(s)	Feedstock condition	Hydrolysis condition	Sugar recoveries	References
Twin-screw extruder	68 °C	6 % NaOH/g d, H₃PO₄ introduced at the end of the extruder	Barley straw, 5 mm	CTec 2 10–40 mg/g substrate, HTec 2 0:9–3:9, 50 °C, 72 h	60–68 % glucose, 52–59 % xylose	Barta et al. (2011)
Single screw extruder l/d 20:1, 3:1 screw compression ratio	45–225 °C, 20–200 rpm	Soaked in 0.5–2.5 % w/w NaOH solution at room temperature for 30 min	Switchgrass, 2–10 mm	15 FPU Celluclast 1.5L and 60 CBU Novo 188/g dm, 50 °C, 72 h, 150 rpm	31–92 % glucose, 41–85 % xylose, 36–86 % combined	Karunanithy and Muthukumarappan (2011f)
Single screw extruder l/d 20:1, 3:1 screw compression ratio	45–225 °C, 20–200 rpm	Soaked in 0.5–2.5 % w/w NaOH solution at room temperature for 30 min	Big bluestem, 2–10 mm	15 FPU Celluclast 1.5L and 60 CBU Novo 188/g dm, 50 °C, 72 h, 150 rpm	33–86 % glucose, 50–94 % xylose, 42–91 % combined	Karunanithy and Muthukumarappan (2011g)
Single screw extruder l/d 20:1, 3:1 screw compression ratio	45–225 °C, 20–200 rpm	Soaked in 0.5–2.5 % w/w NaOH solution at room temperature for 30 min	Prairie cord grass, 2–10 mm	15 FPU Celluclast 1.5L and 60 CBU Novo 188/g dm, 50 °C, 72 h, 150 rpm	37–98 % glucose, 46–95 % xylose, 48–95 % combined	Karunanithy and Muthukumarappan (2011h)
Single screw extruder l/d 20:1, 3:1 screw compression ratio	45–225 °C, 20–200 rpm	Soaked in 0.5–2.5 % w/w NaOH solution at room temperature for 30 min	Corn stover, 2–10 mm	15 FPU Celluclast 1.5L and 60 CBU Novo 188/g dm, 50 °C, 72 h, 150 rpm	25–95 % glucose, 30–90 % xylose, 29–92 % combined	Karunanithy and Muthukumarappan (2011e)

pretreatments without chemicals resulted in a less sugar recovery than that of extrusion pretreatment with chemicals. We performed extrusion pretreatment of different feedstocks with and without alkali soaking and the results are depicted in Fig. 2.5 for easy understanding. The reduction or absence of processing aid led to burning of the feedstock and blocking of the die during extrusion (Lamsal et al. 2010), in order to avoid these problems, Yoo et al. (2011) added starch (5–20 %) and adjusted moisture high range of 20–40 % (wb). If Yoo et al. (2011) would have carried out the extrusion pretreatment without (circular die) as most of the studies listed in Table 2.2, these issues would not have risen due to smooth flow of thefe edstock.

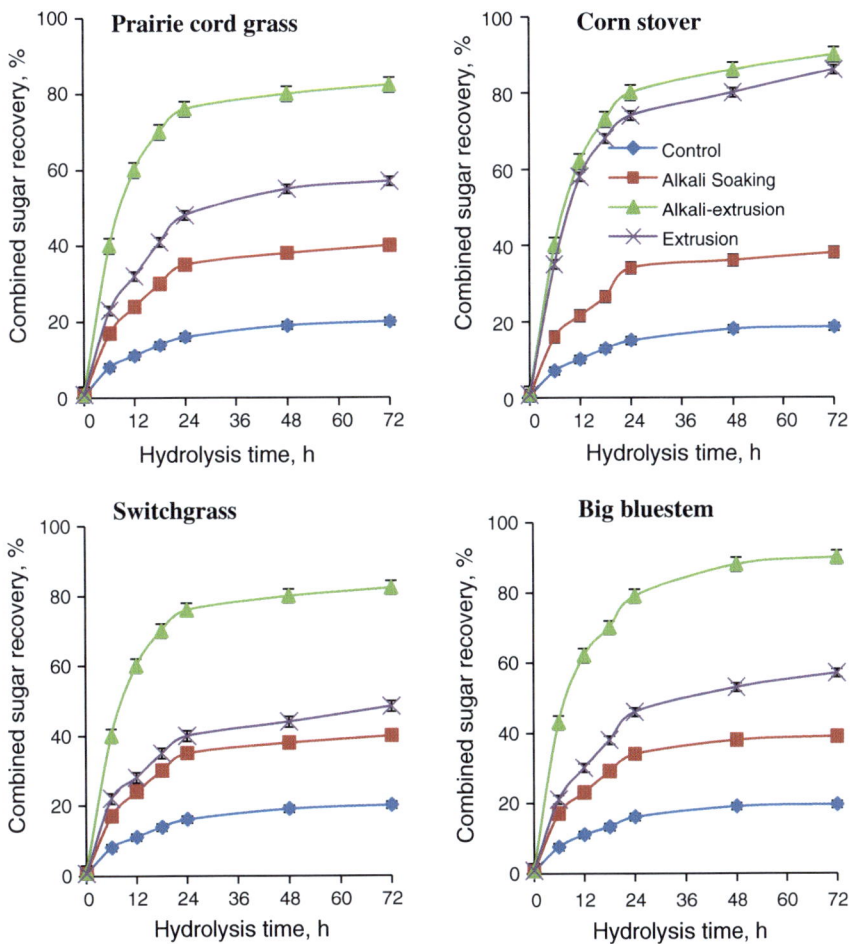

Fig. 2.5 Comparison of sugar recovery profile from control, alkali soaked, and alkali soaked-extruded feedstocks

2.6 EvaluationofPr etreatmentE fficiency

Researchers employed different techniques or methods such as measurement of mean particle size, surface area, pore size and quantity (SEM images), crystallinity index, fermentation inhibitors, and mass balance analysis to support their findings and they are presented in this section.

2.6.1 MeanP articleD iameter

Lee et al. (2009, 2010) observed most of the fibers had a diameter of 20 nm and some in the 100 nm when douglas fir was extruded with cellulose affinity additives, whereas hot compressed water-extrusion resulted in a submicron fiber diameter (5 nm). Different feedstocks pretreated using a single screw extruder is shown in Fig. 2.6 for easy visualization of the changes in particle sizes and they were subjected to particle size analyses (Karunanithy and Muthukumarappan 2011a, b, c, d) The authors reported about 90 % of the control passed through 8 mm and about 95 % of optimum pretreated corn stover, switchgrass, big bluestem, and prairie cord grass passed through 1.78, 1.85, 1.79, and 1.80 mm, respectively. Recently, Chen et al. (2011) found the mean particle diameter of extrusion pretreated rice straw in the range of 400–500 μm depending upon the screw speed, whereas extrusion followed by hot water extraction resulted in lower range of 90–300 μm. The above results provide evidence for increase in sugar recovery from different feedstocks is due to decrease in fiber diameter.

2.6.2 Surface Area

In general, greater the surface area, greater the sugar recovery or hydrolysis yield. Lee et al. (2009) found the specific surface area using two different methods namely, BET method and Congo red dye absorption. They reported the specific surface area as 21.3 and 19.4 m^2/g and amount dye absorbed was 14.3 and 13 mg/g, respectively, for Douglas fir with an ethylene glycol as cellulose affinity additive pretreated at a barrel temperature of 40 and 120 °C in a twin screw extruder. Zhang et al. (2012b) also reported a similar observation i.e., specific surface area of 367–439 m^2/g and amount of dye absorbed was 246–293 mg/g when corn stover pretreated in a twin screw extruder. Karunanithy and Muthukumarappan 2011a, b, c, d) reported that optimum extrusion pretreated corn stover, switchgrass, big bluestem, and prairie cord grass had a surface area of 0.772, 0.688, 0.790, and 1.15 m^2/g when compared to an 8 mm control of 0.400, 0.445, 0.469 and 0.347 m^2/g, respectively, when measured using Beckman Coulter SA3100

Control feedstocks   **Pretreated feedstocks**

Fig.2.6   Feedstocks pretreated in a single screw extruder

surface area analyzer. Recently, Chen et al. (2011) observed when rice straw pretreated with acid in a twin screw extruder had a BET surface area of 2.8–4.0 m²/g

2.6.3 CrystallinityI ndex

Cellulose crystallinity is considered as one of the factors impeding the enzymatic
hydrolysis. Crystalline cellulose offers resistance to enzymatic hydrolysis because of
strong interchain hydrogen-bonding network, whereas amorphous cellulose is readily
digestible, thereby lowering the cellulose crystallinity is desirable. A few researchers have evaluated the crystallinity index of control and extrusion pretreated feedstocks. Though there was an absolute differences among the crystallinity index of
raw Douglas fir, extrusion pretreated at 40 and 120 °C were 68.4, 54.7, and 47.1 %,

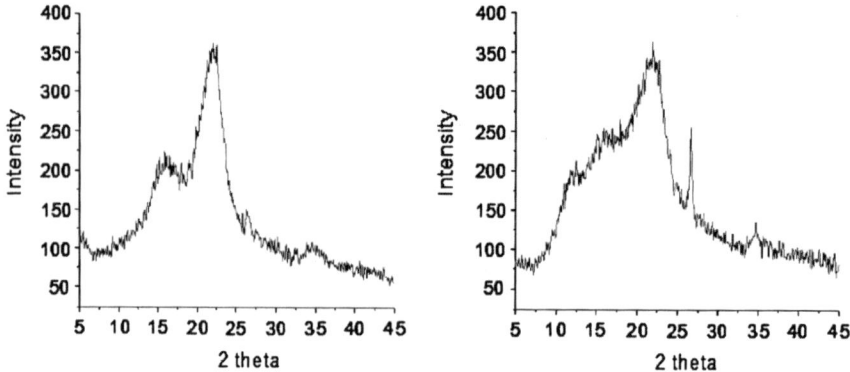

Fig. 2.7 X-ray diffractograms of control and twin-screw extruded at 80 rpm with 27.5 % wb moisture content corn stover (Zhang et al. 2012b). Reprinted from applied biochemistry and biotechnology, pretreatment of corn stover with twin-screw extrusion followed by enzymatic saccharification, © 2011, Shujing Zhang, 2, with kind permission from Springer Science and Business Media and any original (first) copyright notice displayed with material

respectively, they were not significantly different (Lee et al. 2009). Lamsal et al. (2010) noticed that extrusion pretreatment of soybean hull and wheat bran did not lower their crystallinity index. Recently, Zhang et al. (2012b) found that corn stover had a tendency of crystal lattice transition from cellulose I to cellulose II as a result of extrusion pretreatment (Fig. 2.7). The characteristic peaks observed around 12.1° 2θ and a shoulder between 19 and 20° 2θ indicate the presence of cellulose II. However, they did not observe a similar change with other extrusion pretreatment conditions. In another study, Zhang et al. (2012a) observed that alkali extrusion of corn stover had no effect on crystallinity index. The above studies confirm that extrusion facilitates the fibrillation without losing energy to destroy the cellulose crystallinity that raises the question of whether cellulose crystallinity is an impeding factor or not.

2.6.4 PoreSiz eandQ uantity

Some of the researchers have supported their finding with scanning electron microscopic (SEM) images to show the topological changes. One such image is presented for control and extrusion pretreated feedstocks in Fig. 2.8. Senturk-Ozer et al. (2011) have provided the evidence in Fig. 2.9 how the topology changes when mixed wood species was pretreated using reactive extrusion process. One can easily notice the increase in pore size as well as the gradual emergence of separated cellulosic fibers in the extruder upon delignification. Recently, Zhang et al. (2012a) reported an increase in porosity from 73 to 148 mg/g when corn stover extruded depending upon the alkali concentration. However, the authors concluded that higher alkali concentrations not necessarily result in more meso and large-scale pores since the porosity/amount of dye absorbed was not in full agreement with sugar recovery.

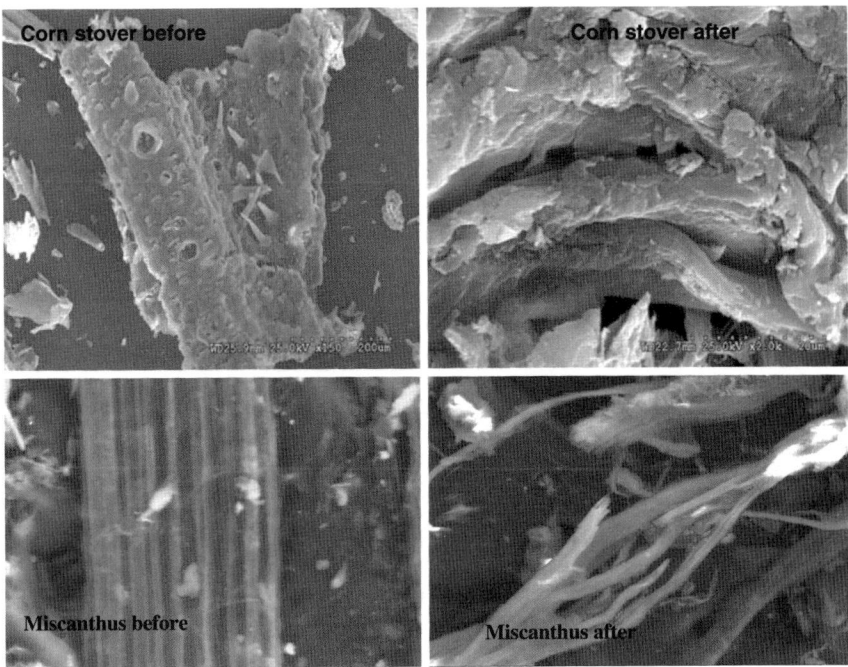

Fig. 2.8 SEM images of corn stover (Zhang et al. 2012b) and miscanthus (Jurisic 2012) before and after extrusion pretreatment. Reprinted from applied biochemistry and biotechnology, pretreatment of corn stover with twin-screw extrusion followed by enzymatic saccharification, © 2011, Shujing Zhang, 2, with kind permission from Springer Science and Business Media and any original (first) copyright notice displayed with material

2.6.5 Washingof Extrudatesis B oonor B ane?

Soybean hull and wheat bran were moisture balanced with sodium hydroxide, urea, and thiourea solution (10 % w/w) to achieve 30–35 and 20–25 %, respectively, and then extruded in a twin screw extruder (Lamsal et al. 2010). Extrudates washed extensively with excess water and centrifugation, HCl was used for adjusting pH to 6.5 and finally the samples were dried at 45 °C for overnight. The reducing sugar of 12–18 % and 38–40 % for soybean hull and wheat bran extrudates, respectively, before washing increased to 28–30 % and 63–70 %, respectively. The authors speculated that extensive washing for the removal of solvents and inhibitors that increased sugar recovery without reporting the amount of solvent and inhibitors details before and after washing, further the effect of HCl addition and pore collapsibility details also missing. They also reported that washing of extrudates made with water alone also improved the reducing sugar without providing sufficient details. According to Yoo et al. (2011), unlike acid or alkali hydrolyzed soybean hulls (Lamsal et al. 2010), the extruded soybean hulls (moisture balanced with water) did not require neutralization or washing steps prior to enzymatic hydrolysis.

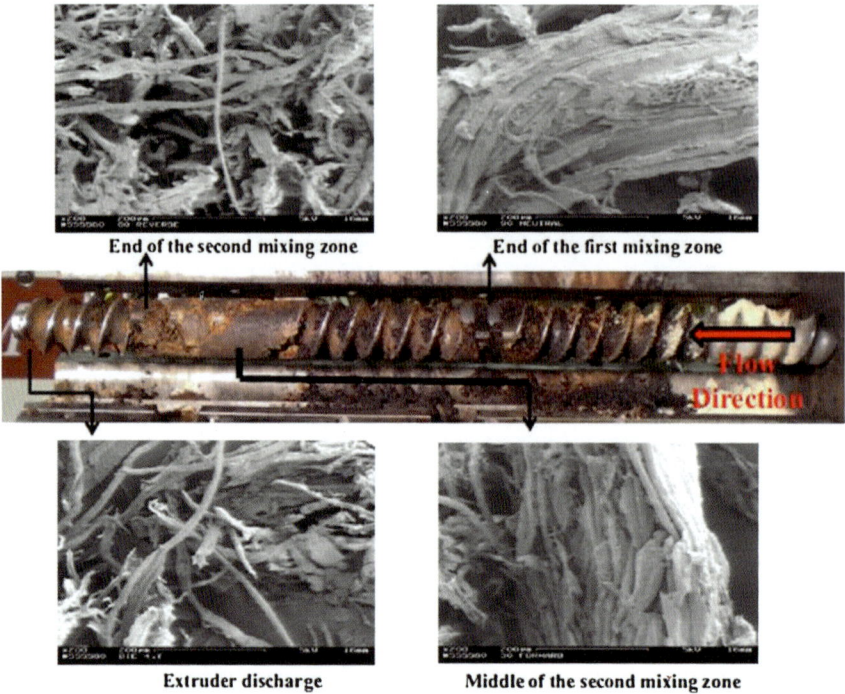

End of the second mixing zone   **End of the first mixing zone**

Extruder discharge   **Middle of the second mixing zone**

Fig. 2.9 SEM micrographs of mixed wood samples collected from different locations in the twin screw extruder (Senturk-Ozer et al. 2011). Reprinted from bioresource technology, biomass pretreatment strategies via control of rheological behavior of biomass suspensions and reactive twin screw extrusion processing, 102/19, 2011, Semra Senturk-Ozer, Halil Gevgilili, Dilhan M. Kalyon, 8, with kind permission from Springer Science and Business Media and any original (first) copyright notice displayed with material

Barta et al. (2011) extruded barely straw with alkali (6 % dm) and then washed the solid fraction with tenfold amount of distilled water. They found the washed solid fraction resulted in higher glucose (60–68 %) and xylose recovery (52–59 %) than that of whole slurry (44–50 % glucose and 40–46 % xylose) at different enzyme loading. Duque et al. (2011) also washed the barely straw extruded with alkali and further details were not found in the conference poster.

Zhang et al. (2012a) extruded corn stover after moisture balancing to 50 % db with alkali solutions (1–14 % w/w), washed with 44 mL of distilled water/g of corn stover extrudates through vacuum filtration and dried the extrudates at ambient conditions for 3 days. They found the extrudates without washing had higher sugar recovery than that of washed extrudates at lower alkali concentrations (1–3 % w/w), but the observation was opposite at higher alkali concentrations (5–14 % w/w). Higher alkali concentration resulted extrudates with high pH that necessitates the washing to neutralize the samples before enzymatic hydrolysis. The authors also attributed pore collapsibility during air drying might outweighed the washing effect at higher alkali concentrations.

The above researchers have reported washing of the feedstocks after extrusion pretreatment; however, it may not be necessary if the alkali concentration is low enough. Zhang et al. (2012a) also found that corn stover extrudates with no washing resulted higher sugar recovery than that of washed extrudates. According to the Novozymes biomass kit, most of the enzymes have optimum activity between a pH of 4.5–6.5 at 45–70 °C. When switchgrass is soaked with different alkali concentrations, the sample prepared for enzymatic hydrolysis (after adding citrate buffer, DI water, and enzymes) had a pH of 4.8–5.4, which is well within the range of Novozyme's recommendations. The authors recorded a sugar recovery of 35–40 % for alkali soaked feedstocks without washing them, depending upon the type of feedstock, alkali concentration, and particle size. It is better to prepare the hydrolysis sample and check the pH before hydrolysis, if the pH is within the recommended pH of the enzymes then there is no need for washing. Moreover, washing removes solubilized sugars (5–7 %) and increases water consumption also.

2.6.6 Fermentationl nhibitors

Compounds like furfural, HMF, acetic acid, and glycerol are formed during feedstock pretreatment depending upon the severities (pretreatment temperature, residence time, and acid concentration); moreover the degradation is proportional to the pretreatment severities. Degradation of pentose and hexose results in formation of furfural and HMF, acetyl content in hemicellulose is hydrolyzed into acetic acid. According to Luo et al. (2002), furfural and HMF inhibits glycolysis and ethanol pathway, protein and RNA synthesis and all these indicate they are potential fermentation inhibitors. Furfural and HMF was not found in any of the extrusion pretreatments employed on different feedstocks with or without alkalis and this makes extrusion pretreatment is better than other leading pretreatments. However, Chen et al. (2011) reported furfural in the range of 0.3–1 g/L when dilute sulfuric acid (1–3 % w/w) added to the rice straw in a twin screw extruder. This indicates that the combination of acidic condition and high temperature leads to furfural formation. Considering this criterion, when sulfuric acid (1.5–3.5 % w/w) added to rape straw pretreated in a twin screw extruder (Choi and Oh 2012) would have resulted in furfural, which was not reported. If the hydrolysate contains furfural in the range of 0.5–8 g/L that would result in an inhibition 21–97 % of ethanol production by *Saccharomyces cerevisiae*, with HMF having a higher threshold level than furfural (Luo et al. 2002; Pienkos and Zhang 2009; Banerjee et al. 1981).

Hydrolysis of the acetyl group present in the hemicellulose is responsible for acetic acid formation as a result of deacetylation. Chemical interference is the mechanism that acetic acid inhibits ethanol production through dissociating in the cytoplasm and causing pH imbalances at high concentration ultimately result in cell growth inhibition or death (Luo et al. 2002). Acetic acid was the most common inhibitors found in low concentrations in the extrusion pretreatments. The hydrolyzate of extruded soybean hull had acetic acid concentration of 1.12–1.42 g/L (Yoo et al. 2012). The authors found acetic acid in the range of 0.04–0.26 g/L for corn stover, 0.03–0.43 g/L for switchgrass (Karunanithy and Muthukumarappan 2011b), 0.05–0.18 g/L for big bluestem (Karunanithy and Muthukumarappan 2012b), and 0.04–0.20 g/L for prairie cord grass (Karunanithy and Muthukumarappan 2011d) when pretreated in a single screw extruder. Similarly, the authors found acetic acid concentration of 0.06–0.17, 0.07–0.16, 0.06–0.1, and 0.07–0.18 g/L, respectively, for corn stover (Karunanithy and Muthukumarappan 2011e), switchgrass grass (Karunanithy and Muthukumarappan 2012b), big bluestem (Karunanithy and Muthukumarappan 2011g), and prairie cord grass (Karunanithy and Muthukumarappan 2011h) when extruded after alkali soaking. However, Karunanithy et al. (2012) reported there was no acetic acid found in the hydrolyzate of extruded pine chips. Chen et al. (Chen et al. 2011) reported higher acetic acid concentration (2.0–5.9 g/L) for rice straw extruded with sulfuric acid than other extrusion studies. If the hydrolyzate contains 1.4–7.5 g/L of acetic acid that would inhibit 50–80 % of ethanol production by *Sacchromyces cerevisiae*; however, the effect is pH dependent (higher inhibition at lower pH) (Olsson and Hanh-Hägerdal 1996; Maiorella et al. 1983; Phowchinda et al. 1995; Pampulha and Loureiro 1989). The concentration of acetic acid found in most of the extrusion studies [extrusion of rice with acid] was far lower than the inhibition limit (5 g/L) (Taherzadeh et al. 1997).

Glycerol is a by-product and it is not directly toxic to yeast. If the glycerol concentration is greater than 100 g/L then it generates osmotic stress in the cells thereby inhibits ethanol fermentation (Maiorella et al. 1983). Extrusion pretreatment of corn stover, switchgrass, big bluestem, and prairie cord grass resulted in a glycerol concentration of 0.02–0.18 g/L (Karunanithy and Muthukumarappan 2010a, b). When these feedstocks subjected to alkali soaking followed by extrusion, a glycerol concentration of 0.04–0.16 g/L was found only for switchgrass (Karunanithy and Muthukumarappan 2012b). Yoo et al. (2012) reported a higher glycerol concentration of 0.24–0.69 g/L for the hydrolyzate of extruded soybean hull. Jurisic (2012) reported byproducts in the range of 0.2–5.1 % on dry matter basis when *miscanthus* pretreated in a single screw extruder. Yoo et al. (2012) reported a significant increase in acetic acid and glycerol concentration between hydrolyzate and fermentation broth of extruded soybean hull. They also found high concentration of lactic acid (7.2–9.3 g/L) only in fermentation broth. Hence, these observations necessitate the fermentation trials of extrusion pretreated feedstocks to ensure extrusion's merit.

2.6.7 MassB alance Analysis

Mass balance provides details on how the pretreatment changes the solid fraction, change in composition of the feedstocks, and degradation/formation of byproducts

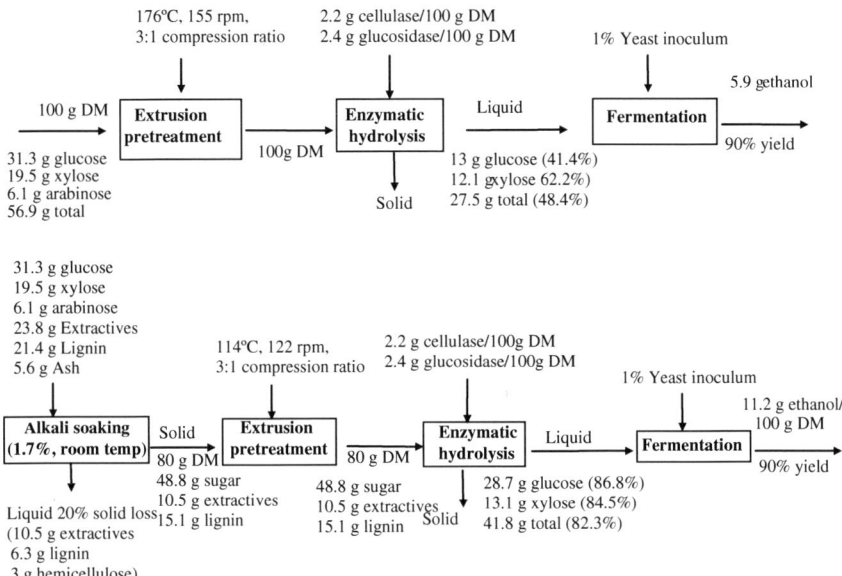

Fig. 2.10 Mass balance diagram of extrusion pretreated switchgrass (Karunanithy andMuthukumarappan 2011b). Reprinted from industrial crops and products, optimization of switchgrass and extruder parameters for enzymatic hydrolysis using response surface methodology, 2011, 33/1, C. Karunanithy, K. Muthukumarappan, 12, with kind permission from Springer Science+Business Media and any original (first) copyright notice displayed with material

if any. Figure 2.10 shows the mass balance of extruded switchgrass grass with and without alkali soaking. As noticed from the figure, alkali soaking resulted in a solid loss of 20 % including small amount of hemicellulose. Preservation of hemicellulose is one of the criteria for an effective pretreatment. The hemicellulose loss can be avoided in the above study if the authors would have used alkali solution for moisture balancing of switchgrass. Mass balance analysis is an important tool for economic reasons; therefore, researchers should report the mass balance analysesofe xtrusion employedondif ferentfe edstocks.

2.7 ComparisonofE xtrusionw ithOthe rPr etreatments

A few researchers have compared different pretreatments with extrusion using the same feedstocks. The authors compared alkali soaking of different feedstocks with extrusion and they found the performance of extrusion (48–57 %) was better than alkali soaking (35–39 %) as shown in Fig. 2.5. However, when alkali soaked feedstocks pretreated in an extruder increased the sugar recovery to 86–95 %. Similarly Zhang et al. (2012a, b) compared extrusion with alkali-extrusion of corn stover and found alkali-extrusion had better glucose (87 %) and xylose (51 %) than that of extrusion alone (49 % glucose and 25 % xylose). A possible reason for increase in sugar recovery is a combination of delignification due to alkali soaking/incorporation and extrusion. The sugar recovery from soybean hull through extrusion pretreatment (155 %) was better than dilute acid (70 %) and alkali (129 %) pretreatments (Yoo et al. 2011, 2012) compared to control. Ng et al. (2011) compared extrusion (50 °C, 100, rpm, 1 straw: 2 water), alkali-extrusion (50 °C, 100 rpm, 4 % NaOH, 1 straw: 2 water), and steam explosion (190 °C, 10 min, 11.6 bars; 220 °C, 10 min, 22.2 bars) of wheat straw. They concluded that low temperature extrusion (32 %) alone yields with 14 % glucose lower in comparison to highest yield from severe stream explosion (46 %). They also reported that alkali-extrusion (42 %) had a comparable glucose yield to severe steam explosion (46 %) and better than steam explosion (40 %). They were in the opinion that temperature would increase the degree of fractionation in extrusion; however, high pretreatment temperature (>200 °C) would generate fermentation inhibitors, thus one has to strike a balance between sugar recovery and inhibitors.

2.8 TorqueandSp ecific Mechanical Energy

The amount of energy required to run the extruder screw is called torque that provides insight into the extruder operation. The motor provides power to turn the screw. Torque relates to the power consumption of the extruder. Torque is related to extruder speed, fill, and viscosity of the material in the screw channel (Harper 1989). Several factors extruder parameters such as screw speed, temperature, compression ratio, and feedstock parameters such as moisture content and particle size are contributing to the net torque. In general, high screw speed and low temperature led to greater torque thereby specific mechanical energy (SME). SME is defined as the mechanical energy input required obtaining unit weight of extrudate. In order to calculate SME, one should know throughput of the feedstock ($\dot{m}$ kg/h).

$$\text{SME (Wh/kg)} = P(W) \times \dot{m}(\text{kg/h}) \tag{2.1}$$

$$P(W) = 220\ V \ \times \ I(A) \times \text{Screw speed (rpm)/Rated screw speed (rpm)} \ \times \ 0.95 \tag{2.2}$$

$$\text{Torque}, \tau \ (\text{Nm}) = 9550 \times \ P \ (\text{kW})/\text{Screw Speed (rpm)} \ \times \ (1/2) \tag{2.3}$$

In P(W) formula, P(W) represents the power consumption, I(A) is the electrical intensity through the driving motor. In torque formula, 9550 is the conversion factor, P(kW) is the measured power, and 1/2 represents the twin-screw (if it is single screw extruder, this term becomes one).

A certain optimum level of torque and SME is necessary to achieve maximum sugar recovery, beyond which negative effects, such as re-condensation of lignin (Karunanithy and Muthukumarappan 2010c) and degradation of sugars (Lamsal et al. 2010; Yoo et al. 2011), might become significant. Lee et al. (2009) found torque as the most effective parameter for fibrillation, compared to the operation temperature and the swelling effect of agents, within the results obtained in this study. They achieved a high torque (~60–75 Nm) at operating temperature of 40 °C in counter rotating twin screw extrusion of Douglas fir led to the highest glucose recovery. The authors compared torque requirement of different feedstocks and correlated with total sugar recovery as depicted in Fig. 2.11. Chen et al. (2011) reported a similar range of torque requirement (107–202 Nm) for extrusion pretreatment of rice straw with acid. Recently, Karunanithy et al. (2012) found the torque requirement as 180–320 Nm for pine chips. In general, the torque requirement/SME is high at low screw speed (Chen et al. 2011; Karunanithy and Muthukumarappan 2011i, 2012a), barrel temperature (Lee et al. 2009; Karunanithy and Muthukumarappan 2011i, 2012a) and feedstock moisture content (Yoo et al. 2012; Karunanithy and Muthukumarappan 2011i, 2012a). However, Yoo et al. (2012) reported an increase in torque with screw speed for soybean hull pretreatment. The authors reported varying optimum ranges of torque requirement during extrusion for maximum sugar recovery from switchgrass (85–100 Nm), corn stover (52–70 Nm), prairie cord grass (27–42 Nm), and big bluestem (53–84 Nm) (Karunanithy and Muthukumarappan 2011i). They also reported the torque requirementde pendsuponthe ligninc ontentofthe fe edstocks.

SME for soybean hull and wheat bran was reported in the range of 566–2615 and 800–2300 kJ/kg, respectively (Lamsal et al. 2010; Yoo et al. 2012). However, addition of chemicals (solution) during extrusion pretreatment of soybean hull and wheat bran reduced the SME 240–540 kJ/kg (Lamsal et al. 2010) due to increase in feedstock moisture content. Rice straw pretreatment in a twin screw extrusion with acid had a SME in the range of 192–496 Wh/kg. The relationship between sugar recovery and torque/SME may not be linear since other factors such as residence time, degree of hydration, degree of fill, and any adverse changes in the substrate would affect the process (Yoo et al. 2012). In order to establish and confirm the relationship between sugar recovery and torque, further investigations are required.

2.9 ExtrusionI nvolvedinSe quentialP retreatments

According to Lee et al. (2010), mechanical kneading alone may not be sufficient for exposing cellulose in submicro/nanoscale for complete enzymatic hydrolysis. With the aim of complete hydrolysis, first Douglas fir and eucalyptus were subjected to hot-compressed water treatment (140–180 °C for 30 min excluding come up time), followed by water washing and then fed to twin screw extruder at room temperature with a screw speed of 45–120 rpm. The glucose yield of sequential pretreated Douglas fir was about 5 times (18–26 wt %) higher than that of hot-compressed water treatment (<5 wt %) alone. Sequential pretreatment resulted in a higher glucose yield for Eucalyptus than for Douglas fir due to the higher hot-compressed water treatment effect in Eucalyptus. Extrusion of Douglas fir with ethylene glycol had a glucose yield of 62.4 % (Lee et al. 2009)

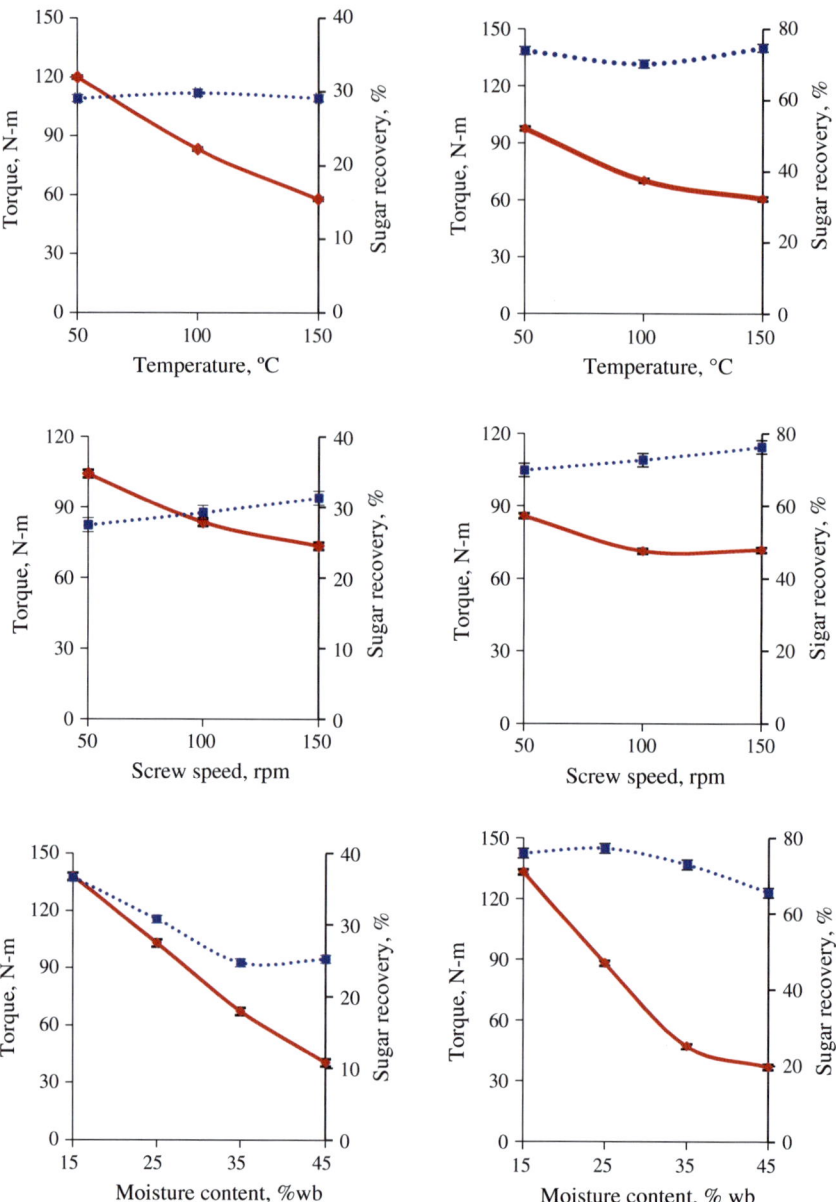

Fig. 2.11 Effect of independent variables on torque requirement (____) and sugar recovery (......) from switchgrass and corn stover (Karunanithy and Muthukumarappan 2012a). Reprinted from BioEnergy research, a comparative study on torque requirement during extrusion pretreatment of different feedstocks, © 2011, Chinnadurai Karunanithy, 5, with kind permission from Springer Science+Business Media and any original (first) copyright notice displayed with material

and this sequential pretreatment had a glucose yield of 67.2 % (Lee et al. 2010). The authors attributed that hot-compressed water treatment facilitated the partial removal of hemicellulose and lignin resulted in loosening the structure in turn the fibrillation (about 20 nm) and hydrolysis. Advantage of this sequential pretreatment is that the extruder can be operated at room temperature with less torque requirement since the feedstock structure was already loosened. Was it worth that the energy spent in the sequential pretreatment for an increase of 5 % glucose yield? Or in other words, is it economical considering only 5 % increase in glucose yield.

In order to increase sugar concentration and overall sugar recovery from rice straw Chen et al. (2011) explored the sequential pretreatment consisting of extrusion with dilute sulfuric acid followed by hot water extraction using saturated steam (130–160 °C) for 10–30 min. They found that enzymatic yield of 19–33 % and 50–61 %, respectively, for only extrusion and sequential extrusion-hot water extraction. The increase in yield was attributed to increase in surface area and decrease in mean particle diameter from 2.8 to 4.0 m^2/g and 400–500 μm (only extrusion)/g to 7–10 m^2/g and 90–300 μm (sequential pretreatment), respectively.

We are exploring different sequential pretreatments consisting microwave or ozone in the front or back end of extrusion and the results are yet to be published. We expect about 20–30 % increase in sugar recovery depending upon the pretreatment conditions.

2.10 FutureD irections

Extrusion pretreatments of different feedstocks clearly demonstrate that it is one of the continuous methods ready for scaling up. However, fermentation results of the extrusion pretreated feedstocks are missing. Fermentation trials of extrusion pretreated soybean hull revealed the presence of lactic acid, though it was not present in the hydrolyzate, but no furfural or HMF presents either in the hydrolyzate or fermentation broth. Hence, in order to prove extrusion pretreatment capability, extruded feedstock should be subjected to fermentation trials to check types of inhibitors and their levels and effects. Though torque requirements of different feedstocks gives fair idea about extrusion pretreatment, energy analysis is missing. Providing complete energy analysis of the extrusion pretreatment of different feedstocks would help the biorefinery industry. The literature survey reveals that extrusion pretreatment alone is capable of getting a sugar recovery in the range of 50–65 %. This indicates sequential pretreatment would be viable option. Considering the cost economics of chemicals and effluent treatment, hot water/ steam pretreatment in the front (preferably) or back end of extrusion would be explored for different feedstocks. Hot water pretreatment in the front end would reduce the torque requirement remarkably during extrusion pretreatment thereby one can save energy cost. In addition, we have indicated several research opportunities in different sections on this chapter.

References

Akdogan H (1996) Pressure, torque, and energy responses of a twin screw extruder at high moisture contents. Food Res Int 29(5–6):423–429

Baillif M, Oksman K (2009) The effect of processing on fiber dispersion, fiber length, and thermal degradation of bleached sulfite cellulose fiber polypropylene composites. J Thermoplast ComposM at22(2):115–133

Banerjee N, Bhatnagar R, Viswanathan L (1981) Inhibition of glycolysis by furfural in *Saccharomycesc erevisiae*.Eur J ApplM icrobiolB iotech11: 226–228

Barta Z, Duque A, Oliva JM, Manzanares P, González A, Ballesteros M, Réczey K (2011) Improving enzymatic hydrolysis of alkali-extruded barley straw by varying the dosage and ratio of hemicellulases and cellulases ITALIC6 Italian meeting on lignocellulosic chemistry final workshop cost FP0602 workshop cost FP0901 5–8 September 2011. Viterbo, Italy

Cadoche L, Lopez GD (1989) Assessment of size reduction as a preliminary step in the production of ethanol from lignocellulosic wastes. Biol Waste 30(2):153–157

Carr ME, Doane WM (1984a) Modification of wheat straw in a high shear mixer. Biotech Bioeng 26:1252–1257

Carr ME, Doane WM (1984b) Pretreatment of wheat straw in a twin screw compounder. Biotech BioengSymp 14:187–195

Chang VS, Kaar WE, Burr B, Holtzapple MT (2001) Simultaneous saccharification and fermentation of lime-treated biomass. Biotechnol Lett 23(16):1327–1333

Chen S, Wayman M (1989) Continuous production of ethanol from aspen cellulose by co-immobilized yeast and enzymes. Process Biochem 24(10):204

Chen FL, Wei YM, Zhang B, Ojokoh AO (2010) System parameters and product properties response of soybean protein extruded at wide moisture range. J Food Eng 96:208–213

Chen W-H, Xu Y-Y, Hwang W-S, Wang J-B (2011) Pretreatment of rice straw using an extrusion/extraction process at bench-scale for producing cellulosic ethanol. Bioresour Technol 102:10451–10458

Choi CH, Oh KK (2012) Application of a continuous twin screw-driven process for dilute acid pretreatmentofra pes traw.B ioresour Technol110: 349–354

Chundawat SPS, Venkatesh B, Dale BE (2007) Effect of particle size based separation of milled corn stover on AFEX pretreatment and enzymatic digestibility. Biotech Bioeng 96(2):219–231

Dale MC, Tyson G, Zhao C, Lei S (1995) The Xylan delignification process for biomass conversion to ethanol. Report DOE/CE/15594-78/Conf-950587-3

Dale BE, Weaver J, Byers FM (1999) Extrusion processing for ammonia fiber explosion (AFEX). ApplB iochemB iotech77: 35–45

de Vrije T, de Haas GG, Tan GB, Keijsers ERP, Claassen PAM (2002) Pretreatment of miscanthus for hydrogen production by *Thermotoga elfi* . Intl J Hydrogen Energy 27(11–12):1381–1390

Desari RK, Bersin RE (2007) The effect of particle size on hydrolysis reaction rates and rheological properties in cellulosic slurries. Appl Biochem Biotech 136–140(1–12):289–299

Draanen AV, Mello S (1997) Production of ethanol from biomass. US Patent No. 5677154

Duque A, Manzanares P, Ballesteros I, Negro MJ, Oliva JM, Gonzalez A, Ballesteros M (2011) Extrusion process for barley straw fractionation

Gibbons WR, Westby CA, Dobbs TL (1986) Intermediate-scale, semicontinuous solid phase fermentation process for production of fuel ethanol from sweet sorghum. Appl Environ Microbiol51(1):1 15–122

Gould JM (1984) Alkaline peroxide delignification of agricultural residues to enhance enzymatic saccharification.B iotechB ioeng26: 46–52

Gould JM (1985) Enhanced polysaccharide recovery from agricultural residues and perennial grasses treated with alkaline peroxide. Biotech Bioeng2 7:893–896

Gould JM, Jasberg BK, Fahey GC Jr, Berger LL (1989) Treatment of wheat straw with alkaline hydrogen peroxide in a modified extruder. Biotech Bioeng 33(2):233–235

Guo G-L, Chen W-H, Chen W-H, Men L-C, Hwang W-S (2008) Characterization of dilute acid pretreatment of silvergrassf ore thanolpr oduction.B ioresour Technol99(14):6046–6053

Harper JM (1981) Extrusion of foods, vol. I. CRC Press, Inc., Boca Raton. IBSN 0-8493-5203-3

Harper JM (1989) Food extruders and their applications. In Mercier C, Linko P, Harper JM (eds) Extrusion cooking. American Association of Cereal Chemists, St. Paul, p 1e15

Hayashi N, Hayakawa I, Fujio Y (1992) Hydration of heat-treated soy protein isolate and its effect on the molten flow properties at an elevated temperature. Intl J Food Sci Tech 27(5):565–571

Helming O, Arnold G, Rzehak H, Fahey GC Jr, Berger LL, Merchen NR (1989) Improving the nutritive value of lignocellulosics: the synergistic effects between alkaline hydrogen peroxide and extrusion treatments. Biotech Bioeng33: 237–241

Hu Z, Wang Y, Wen Z (2008) Alkali (NaOH) pretreatment of switchgrass by radio frequency-based dielectric heating. Appl Biochem Biotech 148(1–3):71–81

Jurisic V (2012) Optimization of high shear extrusion pretreatment of grass from genus *miscanthus* as raw material in bioethanol production. PhD thesis, submitted to University of Zagreb,C rotia

Jurisic V, Karunanithy C, Julson JL (2009) Effect of extrusion pretreatment on enzymatic hydrolysis of Miscanthus. ASABE Paperno.097178.St .J oseph,M ich: ASABE

Karunanithy C, Muthukumarappan K (2010a) Effect of extruder parameters and moisture content of corn stover and big bluestem on sugar recovery. Biol Eng 2(2):91–113

Karunanithy C, Muthukumarappan K (2010b) Effect of extruder parameters and moisture content of switchgrass, prairie cord grass on sugar recovery from enzymatic hydrolysis. Appl Biochem Biotech 162:1785–1803

Karunanithy C, Muthukumarappan K (2010c) Influence of extruder temperature and screw speed on sugar recovery from corn stover through enzymatic hydrolysis while varying enzymes and their ratios. Appl Biochem Biotech 162:264–279

Karunanithy C, Muthukumarappan K (2011a) Optimization of corn stover and extruder parameters for enzymatic hydrolysis using response surface methodology. Biol Eng 3(2):73–95

Karunanithy C, Muthukumarappan K (2011b) Optimization of switchgrass and extruder parameters for enzymatic hydrolysis using response surface methodology. Ind Crops Prod 33(1):188–199

Karunanithy C, Muthukumarappan K (2011c) Optimization of big bluestem and extruder parameters for enzymatic hydrolysis using response surface methodology. Inter J Agric Biol Eng 4(1):61–74

Karunanithy C, Muthukumarappan K (2011d) Optimization of extruder and prairie cord grass parameters for maximum sugar recovery through enzymatic hydrolysis. J Biobased Mat Bioenergy5(4):520–531

Karunanithy C, Muthukumarappan K (2011e) Chapter 14: application of response surface methodology to optimize the alkali concentration, corn stover particle size and extruder parameters for maximum sugar recovery. In: MA Dos Santos Bernardes (ed) Biofuel production-recent development and prospects, pp 343–374. ISBN:978-953-307-478-8

Karunanithy C, Muthukumarappan K (2011f) Optimization of alkali concentration, switchgrass particle size and extruder parameters for maximum sugar recovery using response surface methodology.C hemEng Tech34(9):1413–1426

Karunanithy C, Muthukumarappan K (2011g) Optimization of alkali concentration, big bluestem particle size and extruder parameters for maximum sugar recovery using response surface methodology. BioResources 6(1):762–790

Karunanithy C, Muthukumarappan K (2011h) Optimization of alkali concentration and extruder parameters for maximum sugar recovery from prairie cord grass using response surface methodology. Biochem Eng J 54:71–82

Karunanithy C, Muthukumarappan K (2011i) Influence of extruder and biomass variables on torque requirement during pretreatment of different biomasses—a response surface analysis. BiosystEng 109(1) :37–51

Karunanithy C, Muthukumarappan K (2012a) A comparative study of torque requirement during extrusion pretreatment for different biomasses. BioEnergy Res 5:265–276

Karunanithy C, Muthukumarappan K (2012b) Extrusion pretreatment of biomass towards bioethanol production. LAP LAMBERT Academic Publishing GmbH & Co, Saarbrücken. ISBN 978-3-8473-4262-5

Karunanithy C, Muthukumarappan K, Gibbons WR (2012) Extrusion pretreatment of pine wood chips. ApplB iochemB iotechnol.doi :10.1007/s12010-012-9662-3

Kerley MS, Fahey GC Jr, Berger LL, Gould JM, Baker FL (1985) Alkaline hydrogen peroxide treatment unlocks energy in agricultural by-products. Science 203:820–822

Kerley MS, Fahey GC Jr, Berger LL, Gould JL, Merchen NR, Gould JM (1986) Effects of alkaline hydrogen peroxide treatment of wheat straw on site and extent of digestion in sheep. J AnimalSc i63:868 –878

Kim SB, Lee YY (2002) Diffusion of sulfuric acid within lignocellulosic biomass particles and its impact on dilute acid treatment. Bioresour Technol83(2):165–171

Lamsal B, Yoo J, Brijwani K, Alavi S (2010) Extrusion as a thermo-mechanical pre-treatment for lignocellulosic ethanol. Biomass Bioenergy34: 1703–1710

Laser M, Schulman D, Allen SG, Lichwa J, Antal MJ Jr, Lynd LR (2002) A comparison of liquid hot water and steam pretreatments of sugarcane bagasse for bioconversion to ethanol. Bioresour Technol81(1):33–44

Lee SH, Teramoto Y, Endo T (2009) Enzymatic saccharification of woody biomass micro/nanofibrillated by continuous extrusion process: I: effect of additives with cellulose affinity.B ioresour Technol100(1):275–279

Lee SH, Inoue S, Teramoto Y, Endo T (2010) Enzymatic saccharification of woody biomass micro/nanofibrillated by continuous extrusion process II: effect of hot-compressed water treatment.B ioresour Technol101: 9645–9649

Lewis SM, Kerley MS, Fahey GC Jr, Berger LL, Gould JM (1987) Use of alkaline hydrogen peroxide-treated wheat straw as an energy source for the growing ruminant. Nutr Rep Int 35:1093–1104

Luo C, Brink D, Blanch W (2002) Identification of potential fermentation inhibitors in conversion of hybrid poplar hydrolyzate to ethanol. Biomass Bioenergetics 22:125–138

Maiorella B, Blanch HW, Wilke CR (1983) By-product inhibition effects on ethanolic fermentation by Saccharomycesc erevisiae. Biotechnol Bioeng25: 103–121

Miller S, Hester R (2007) Concentrated acid conversion of pine softwood to sugars Part I: Use of a twin-screw reactor for hydrolysis pretreatment. Chem Eng Comm 194(1):85–102

N'Diaye S, Rigal L, Larocque P, Vidal PF (1996) Extraction of hemicelluloses from poplar, Populus tremuloides, using an extruder-type twin screw reactor: a feasibility study. Bioresour Technol57 :61–67

National Research Council (2000) Biobased industrial products—priorities of research and commercialization. National Academy Press, Washington, DC

Ng TH, Song J, Tarverdi K (2011) Cereal straw pre-treatment for bio-fuel application: comparison between extrusion and conventional steam explosion In: ResCon'11, 4th annual research student conference, pp 14–17, 20–22 June 2011

Noon R, Hochstetler T (1982) Production of alcohol fuels via acid hydrolysis extrusion technology. Fuel Alcohol U S A 4(7):14–23

Olsson L, Hanh-Hägerdal B (1996) Fermentation of lignocellulosic hydrolysates for ethanol production.Enz ymM icrob Tech18: 312–331

Pampulha ME, Loureiro V (1989) Interaction of the effect of acetic acid and ethanol on inhibition of fermentation in Saccharomycesc erevisiae. Biotechnol Lett 11:269–274

Phowchinda O, Delia-Dupuy ML, Strehaiano P (1995) Effect of acetic acid on growth and fermentation activity of Saccharomycesc erevisiae. Biotechnol Lett 17:237–242

Pienkos PT, Zhang M (2009) Role of pretreatment and conditioning processes on toxicity of lignocellulosic biomass hydrolysates. Cellulose 16:743–762

Senturk-Ozer S, Gevgilili H, Kalyon DM (2011) Biomass pretreatment strategies via control of rheological behavior of biomass suspensions and reactive twin screw extrusion processing. Bioresour Technol102: 9068–9075

Taherzadeh MJ, Niklasson C, Lidén G (1997) Acetic acid—friend or foe in anaerobic batch conversion of glucose to ethanol by *Saccharomyces cerevisiae*? Chem Eng Sci 52(15):2653–2659

Tyson G (1993) Evaluation of alternate pretreatment and biomass fractionation process. NREL report on contract XAW-3-11181-05-104840

Williams BA, van der Poel AFB, Boer H, Tamminga S (1997) The effect of extrusion conditions on the fermentability of wheat straw and corn silage. J Sci Food Agric 74(1):117–124

Yang B, Wyman CE (2008) Pretreatment: the key to unlocking low-cost cellulosic ethanol. Biofuels,B ioprodB iorefin2: 26–40

Yang C, Shen Z, Yu G, Wang J (2008) Effect and aftereffect of -radiation pretreatment on enzymatic hydrolysis of wheat straw. Bioresour Technol 99(14):6240–6245

Yeh A-I, Jaw Y-M (1998) Modeling residence time distributions for single screw extrusion process. J Food Eng 35(2):211–232

Yoo J, Alavi S, Vadlani P, Vincent AB (2011) Thermo-mechanical extrusion pretreatment for conversion of soybean hulls to fermentable sugars. Bioresour Technol 102:7583–7590

Yoc J, Alavi S, Vadlani P, Behnke KC (2012) Soybean hulls pretreated using thermo-mechanical extrusion—hydrolysis efficiency, fermentation inhibitors, and ethanol yield. Appl Biochem Biotechnol.doi:10.1007/ s12010-011-9449-y

Zhang S, Keshwani DR, Xu Y, Hanna MA (2012a) Alkali combined extrusion pretreatment of corn stover to enhance enzyme saccharification. Ind Crops Prod 37:352–357

Zhang S, Xu Y, Hanna MA (2012b) Pretreatment of corn stover with twin-screw extrusion followed by enzymatic saccharification. ApplB iochemB iotechnol166: 458–469

Chapter 3
Solid-State Biological Pretreatment of Lignocellulosic Biomass

Caixia Wan and Yebo Li

Abstract Interest in biological pretreatment of lignocellulosic biomass has shifted from traditional applications, such as ruminant feed upgrading and biopulping, to biofuel production. Biological pretreatment is considered to be a "green" technology as it is performed under ambient conditions without chemical addition. The main benefits include low energy requirements and little or no waste stream output. It has the potential to be applied to on-farm wet storage for cost-effective biofuels production from lignocellulosic biomass. White rot fungi are particularly suitable for biological pretreatment as they enzymatically degrade lignin through their unique ligninolytic systems. This chapter reviews biological pretreatment of lignocellulosic biomass with white rot fungi under solid-sate fermentation for on-farm application. The topics discussed focus on ligninolytic systems, processing conditions, and degradation effectiveness of lignocellulosic biomass.

Keywords Biological pretreatment • Lignocellulosic biomass • Wet storage • Biofuels

3.1 Introduction

One of the greatest challenges of producing biofuels from lignocellulosic biomass is to reduce biomass recalcitrance to hydrolytic microorganisms or enzymes. To accomplish this, various pretreatment methods have been used to break down the complex lignocellulosic biomass and, consequently, facilitate its conversion

C. Wan
Department of Environmental Science and Engineering,
Fudan University, 200433 Shanghai, China

Y. Li (✉)
Department of Food, Agricultural, and Biological Engineering,
Ohio State University/OARDC, Wooster, OH 44691, USA
e-mail: li.851@osu.edu

T. Gu (ed.), *Green Biomass Pretreatment for Biofuels Production*,
SpringerBriefs in Green Chemistry for Sustainability,
DOI: 10.1007/978-94-007-6052-3_3, © The Author(s) 2013

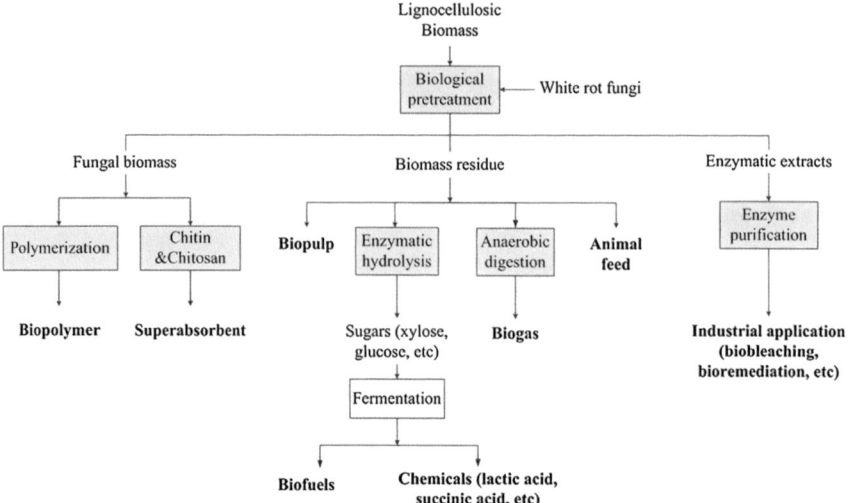

Fig. 3.1 Suggested biological pretreatments with white rot fungi for various applications (Diagramm odified from Isroi et al. 2011)

into biofuels and bioproducts. Thermochemical methods, such as dilute acid and ammonia explosion, can efficiently fractionate biomass feedstocks and thus are considered the current leading pretreatment technologies (Moiser et al. 2005). However, inherent drawbacks with these methods, such as high capital cost, intensive energy requirement, and waste stream generation, have become barriers to their industrial applications. Biological pretreatment, which generally involves lignin-degrading organisms, is a low carbon-footprint technology and can be an alternative to thermochemical pretreatment in many applications. Figure 3.1 summarizes applications of biological pretreatment of lignocellulosic biomass for biobased products and biofuels. Traditional biological pretreatment applications are biopulping and ruminant feed upgrading, while recent studies have focused on biogas,b io-oil,a ndb iofuelp roduction.

Microorganisms that secrete multiple cell wall degrading enzymes have been effectively used for biological pretreatment. These microorganisms include wood rot fungi, ruminant bacteria, and symbiotic bacteria found in some invertebrate animals (e.g., termites, earthworms). Wood rot fungi (e.g., white rot, brown rot, and soft rot) can degrade or modify lignin to some extent through ligninolytic enzymes (Eriksson et al. 1990). Symbiotic microbes in animal rumens or digestive tracts of termites can produce cellulolytic and hemicellulolytic systems, which are largely responsible for hydrolysis of the biomass taken in by their hosts (Varm et al. 1994). Invertebrate animals, like termites and earthworms, are also believed to produce highly active cellulase by themselves (Watanabe and Tokuda 2001). Nevertheless, wood decay fungi, especially white rot fungi, are the most attractive and widely studied for biological pretreatment due to their unique ligninolytic systems (e.g., lignin peroxidase, laccase, and manganese peroxidase).

Biological pretreatment is mostly conducted via a solid-state fermentation (SSF) process. Compared to thermochemical pretreatments, biological pretreatment takes a longer time as the microbes, especially fungi, slowly colonize and decompose biomass feedstocks. This makes this technology less feasible for on-site pretreatment of lignocellulosic biomass in a processing plant. To solve this problem, Wan and Li (2010a, b) have proposed applying a fungal pretreatment to wet storage of agricultural residues on farms. Such emerging in-storage pretreatment generally incorporates long pretreatment times and thus can successfully provide year-round delignified biomass to biorefineries (Digman et al. 2010; Shinner et al. 2007). Also controlling and optimizing influential factors can speed up fungal degradation. Another issue with microbial pretreatment is consumption of carbohydrates by microbes for self-growth and metabolism. Therefore, microbes which selectively degrade lignin over cellulose are preferred for biological pretreatment. Despite these disadvantages, biological pretreatment has great potential to reduce environmental impacts and energy expenditure relative to current prevailing pretreatment technologies. Furthermore, compared to ensilages that harness lactic acid bacteria for fermentation, fungal pretreatment applied as an on-farm wet storage pretreatment would provide pretreated biomass residues with much higher cellulose digestibility.

3.2 WhiteR otFungi

Wood rot fungi are the most understood lignin degrading microorganisms and belong to the ascomycete or basidiomycete groups. According to their decay patterns, wood rot fungi can be classified into three categories: white rot, brown rot, and soft rot. Among these, white rot fungi are the only group that can completely degrade lignin to CO_2 and H_2O, and thus have received extensive interest for delignifying lignocellulosic biomass (Kirk and Farrell 1987). The most widely studied white rot fungi include *Phanerochaete chrysosporium, Ceriporiosis subvermispora, Cyathus stercoreus, Dichomitus squalens, Phlebia radiate, Pleurotlls ostreatus,* and *Trametes versicolor*. During white rot decay, fungal hyphae massively colonize the ray parenchyma cells of the biomass feedstock, then penetrate through pits and form numerous boreholes and erosion troughs on the vessel beneath or around the hyphae (Fig. 3.2). At the advanced stage of decay, the feedstock generally changes to a whitish-yellow color and becomes light, soft, and spongy. Reactions within the lignin by this group of fungi include side-chain oxidation, propyl side-chain cleavage, and demethylation (Chen et al. 1983). Although white rot fungi attack polysaccharides, some of them are able to degrade hemicellulose rather than cellulose. Such selective degradation leaves delignified and cellulose-rich residues for subsequent processing (Blanchette 1995). Therefore, white rot fungi with high selectivity are important for biological pretreatment. Genetic engineering has been used to modify some fast-growing but nonselective white rot fungi, like *P. chrysosporium*, in order to

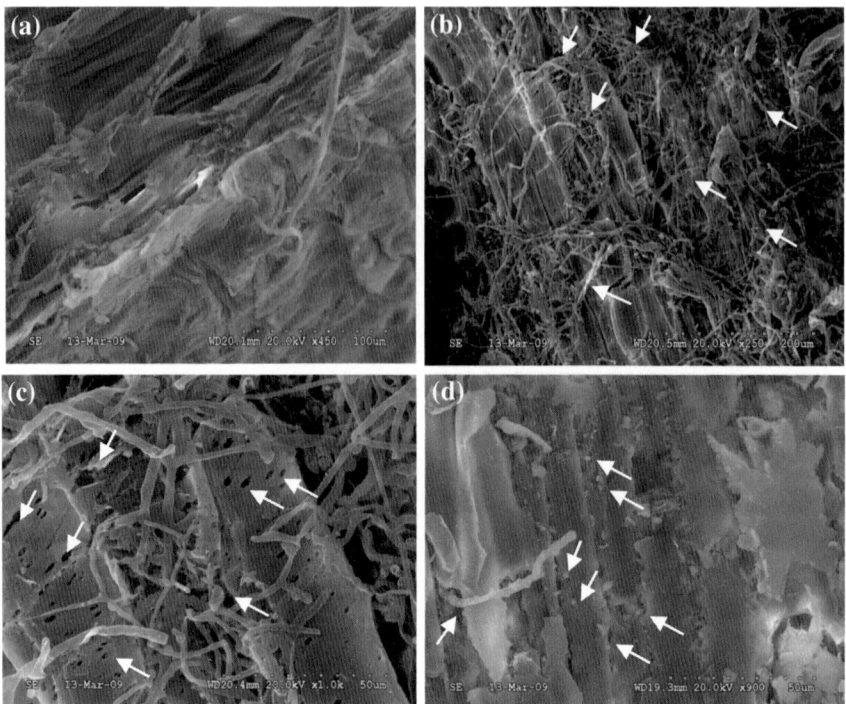

Fig. 3.2 Colonization of fungal hyphae on lignocellulose biomass (images from Wan 2011). Scanning electron micrograph of corn stover pretreated by *C. subvermispora* for 18 d. **a** untreated corn stover (at 450 × magnification), **b** fungal-pretreated corn stover (at 250 × magnification) showing numerous erosion troughs and holes on the vessel wall (*arrows*), **c** fungal pretreated corn stover (at 1000 × magnification) showing cracks and cavities with erosion troughs enlarged (*arrows*), and **d** fungal pretreated corn stover (at 900 × magnification) showing hyphae penetration through cell walls (*arrows*)

obtain cellulase-deficient strains (Eriksson et al. 1983). Unfortunately, reports on the performance and degradation efficiency of lignocellulosic biomass by genetically modified strains have not been as good as expected.

Unlike white rot fungi, brown rot fungi preferentially degrade polysaccharides but only modify lignin to a limited extent (Green and Highley 1997). These fungi modify lignin via demethylation or hydroxylation but with no fragmentation of the aromatic ring (Kirk and Highley 1973). Important brown rot fungi include *Tyromyces balsemeus, Gloeophyllum trabeum, Poria placenta, Lentinus lepidius, Lenzites trabeum, Coniophora puteana, Fomitopsis pinicola, Laetiporus sulfureus,* and, similar to brown rot fungi, soft rot fungi (e.g., *Chaetomium cellulolyticum, Aspergillus niger, Thielavia terrestrisorllm*) also efficiently degrade polysaccharides while slowly and slightly altering the lignin structure. Wood decayed by soft rot fungi is depleted with carbohydrates but rich in lignin, with a noticeably softteneds urface(B lanchette1995).

3.3 LigninolyticSys tems

Lignin, a complex phenolpropanoid polymer, provides structural support for plants and also serves as a natural barrier to microorganism attack on plant tissues. Lignin primarily arises from three hydroxycinnamyl alcohol precursors that differ in their degree of methoxylation: *p*-coumary, coniferul, and sinapyl alcohols (Campbell and Sederoff 1996). These alcohols are randomly copolymerized through one-electron oxidation, which contributes to the extremely heterogeneous and highly rigid structure of lignin. Due to these structural features, biodegradation of lignin must be accomplished through oxidative mechanisms by extracellular and nonspecific enzyme systems (Kirk and Farrell 1987). Aerobic conditions are essential for oxidative cleavage of lignin subunits, specifically carbon–carbon and ether bonds. Primary lignin-degrading enzymes include lignin peroxidase (LiP, EC 1.11.1.14), manganese peroxidase (MnP, EC 1.10.1.13), and laccase (Lac, EC 1.11.1.13). Versatile peroxidase (VP, EC 1.11.1.16) also has an important role in lignin degradation. In addition, some accessory enzymes, such as aryl alcohol oxidase (AAO, EC 1.1.3.7) and glyoxal oxidase (GLOX), are involved in H_2O_2 generation, which provide extracellular H_2O_2 for oxidative turnover of MnPs and LiPs. White rot fungi possess gene families that encode the enzymes responsible for oxidation of lignin and its structural analogues (Hatakka 1994). However, not all of the above enzymes are found in a single fungal culture. For example, LiP activity was not detected in the extensively studied white rot fungus, *C. subvermispora*, although a *lip*-like gene was revealed in this fungus (Rajakumar et al. 1996). The key enzymes responsible for lignin depolymerization are discussed below.

3.3.1 LigninP eroxidase

Lignin peroxidase (LiP, 1.11.1.14) was first found in a liquid culture of *P. chrysosporium* and later in other white rot fungi, e.g., *Phlebia radiate*, *Coriolus versicolor*, *Pleurotus ostreauts*, *T. versicolar*, *Bjekandera* sp., and *T. cervina*. LiPs are heme-containing proteins in which the iron is presented as Fe^{3+} in a porphyrin ring (Farrcll ct al. 1989, Wong 2009). The molecular weight of LiPs is around 40 kDa. Figure 3.3 illustrates the mechanism of LiP-catalyzed oxidation of lignin polymers (Breen and Singleton 1999). Initially, LiPs are oxidized by H_2O_2 to form an intermediate (compound I) that is deficient in two electrons. Compound I then extracts one electron from the donor substrate, resulting in a reduced intermediate (compound II) and a cation radical intermediate. Compound II can in turn oxidize a second molecule of the donor substrate, also through one electron transfer, yielding a cation radical intermediate and the resting state of the peroxidase. These cation radical intermediates spontaneously break into small fragments. In the absence of a suitable donor substrate, compound II can be further oxidized to compound III by H_2O_2 (Dosoretz et al. 2004). Compound III readily returns to the native LiP state if H_2O_2 and veratryl alcohol (VA) are present. However, excessive

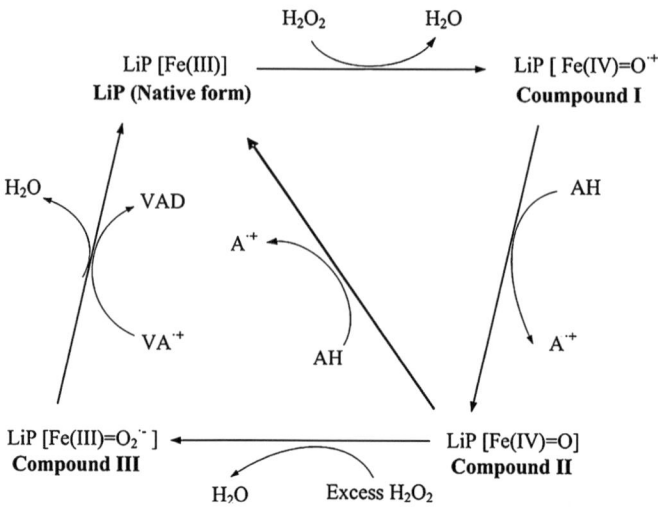

Fig. 3.3 Oxidative mechanism of lignin peroxidase (LiP). AH is an electron-donor substrate; A^+ denotes a cation radical; VA^+ denotes VA radical cation. LiP is activated by H_2O_2 to form intermediates capable of oxidizing lignin-related compounds following one electron transfer. (Modified from Applied Biochemistry and Biotechnology, 157/2, Structure and Action Mechanism of Ligninolytic Enzymes, Dominic W. S. Wong, 1, Copyright 2008, with kind permission of Springer Science + Business Media)

H_2O_2 inactivates compound III. VA plays an important role in protecting LiP from so-called H_2O_2-dependent inactivation by providing the intermediate VA^+ that is capableofre ducingc ompoundIII toits na tives tate.

LiPs are strongly oxidizing, preferentially cleaving C_α–C_β bonds in lignin-related compounds (Hammel et al. 1986). Compared to classical peroxidases, such as enzymes from horseradish, LiPs are capable of oxidizing bulky lignin-related substrates (e.g., polycyclic aromatic hydrocarbons phenols, aromatic amines) (Breen and Singleton 1999). The highly oxidative activity of LiPs could be due to the electron deficiency of the iron on the porphyrin ring (Millis et al. 1989). In addition, invariant tryptophan residue in LiP proteins, which is believed to have important functions for electron transfer from aromatic compounds, presents on enzyme surface with sufficient exposure and thus enables LiPs to directly oxidize nonphenolic lignin-related structures (Doyle et al. 1998; Hammel and Cullen 2008). Veratryl alcohol, a secondary metabolism produced by some white rot fungi, has been postulated to function as an efficient reducing agent to counter possible oxidative inactivation of LiPs caused by H_2O_2 (Hammel et al. 1993). For larger lignin models, in particular lignocellulosic biomass, LiPs are too large of a molecule to diffuse into sound cell walls (Blanchette et al. 1997; Srebotnik et al. 1988). However, these oxidative enzymes could induce the formation of diffusible radicals at the surface of the cell wall (Enoki et al. 1999; Kapich et al. 1999). These radicals are diffusible and could initiate biomass decay and facilitate the penetration of lignin-degrading enzymes (Galkin et al. 1998). Cation radicals of veratryl alcohol can act as such a mediator, which also explains the effect of veratryl alcohol on LiP-catalyzed reactions (Gilardi et al. 1990). However, there is no strong evidence supporting the proposed mechanism that LiPs act on such indirect lignin oxidation.

There are ten LiP isoenzymes, termed as LiPA to LiPJ, found in *P. chrysosporium*, but they are similar in reactivity (Gaskell et al. 1994). Expression of LiP isoenzyme genes is differentially regulated by culture conditions and also varies from strain to strain. In chemical defined media, transcripts lip*A*, lip*C*, and lip*J* were sufficiently upregulated under nitrogen deficiency while the situation was reversed under carbon deficiency. In addition, transcript expression patterns and levels were impacted by substrate sources. For example, lip*D* and lip*E* transcripts were highly expressed in soil but not in aspen wood (Bogan et al. 1996; Janse et al. 1998). However, no similarity of transcript expression profiles has been observed in chemical-defined media or in complex substrates such as wood.

3.3.2 Manganese Peroxidase

Most white rot fungi produce a manganese peroxidase system (MnP, EC 1.11.1.7). Similar to LiPs, MnPs are heme-containing glycoproteins that functionally require H_2O_2 (Kirk and Farrell 1987). MnPs are also strong oxidants, capable of oxidizing numerous lignin-related compounds (Gold et al. 2000). Unlike LiPs, MnPs generate a large amount of small and diffusible oxidizing agents and in turn oxidize lignin indirectly (Breen and Singleton 1999). Figure 3.4 shows proposed oxidative mechanisms of MnP (Kirk and Cullen 1998). As with manganese binding sites,

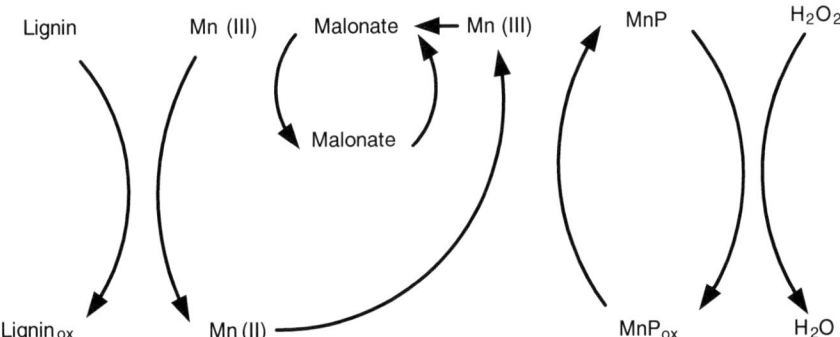

Fig. 3.4 Oxidative mechanisms of Manganese Peroxidase (MnP). OX denotes the oxidized state. MnPs oxidize Mn^{2+} by one electron to Mn^{3+}. Mn^{3+} can be released from active binding sites as a form of Mn^{3+}-organic acid chelates which are capable of oxidizing lignin substrates (phenolics). (Reprinted from Current Opinion in Biotechnology, 10/3, Alec Breen, Fred L Singleton, Fungi in lignocellulose breakdown and biopulping, 7, Copyright 1999, with permission from Elsevier)

MnPs oxidize Mn^{2+} by one electron to Mn^{3+}, which can be released from active binding sites as a form of Mn^{3+}-organic acid chelates. Oxalate, a secondary metabolite of many white rot fungi, is generally thought to act as the chelator to stabilize Mn^{3+}. On the other hand, stabilizing considerably decreases the oxidizing power of Mn^{3+} and could result in limited lignin oxidation. In light of MnPs highly oxidative activity, other oxidants, like peroxyl and acyl radicals, generated from subsequent reaction of Mn^{3+} should be able to extensively degrade lignin (Kapich et al. 1999; Watanabe et al. 2000). It has been shown that peroxyl radicals generated from lipid peroxidation catalyzed by MnPs in the presence of chelated Mn^{3+} and H_2O_2 are capable of cleaving nonphenolic lignin structures (Kirk and Cullen 1998).

MnP isoenzymes, as many as 11 in *C. subvermisporo*, are similar to LiPs in that they do not differ from each other in reactivity (Breen and Singleton 1999). However, it is still unclear why so much redundancy exists in both peroxidase enzymes (Hammel and Cullen 2008). Relative to LiP gene expression, MnP is more differentially regulated not only by culture conditions, such as nutrient limitation and Mn^{2+} concentration, but also by other physiological factors (e.g., temperature, agitation, moisture) (Janse et al. 1998). Identification of regulatory elements in promoter regions indicates that the expression of genes encoding ligninolytic enzymes is regulated at the transcription level. Activator protein-2-binding sequences in the upstream of *lip* and *mnp* genes could respond to nitrogen limitation, while cAMP response elements (CRE) could have a role in either carbon or nitrogen regulation (Dhawale 1993; Have and Teunissen 2001). Putative metal response elements (MREs), responding to heavy metals regulation, have been identified in the upstream regulatory region of *mnp1* and *mnp2* genes but not in that of *mnp3*, which corresponds to high dependency of *mnp1* and *mnp2* on Mn^{2+} with no *mnp3* response to Mn^{2+} (Breen and Singleton 1999). Some regulatory elements that respond to physiological factors, such as *cis*-regulatory element's response to heat shock, are also identified upstream of *mnp1* in *P.c hrysosporium*.

3.3.3 Laccase

Laccases (Lac, EC 1.10.3.2) are widespread and can be found in fungi, bacteria, and plants. Most white rot fungi secret laccase but *P. chrysosporium* apparently lacks detectable laccase activity even with putative *lcc* genes (Bollag and Leonwicz 1984). Laccases are blue-copper containing phenoloxidases with molecular masses between 60 and 80 kDa, capable of oxidizing phenolics and similar molecules by one electron (Gianfreda et al. 1999). The oxidative mechanism of laccase on phenolic lignin-related molecules is proposed as follows: the laccase removes one electron from a phenolic nucleus, generating phenoxyl radicals, which leads to polymer cleavage that mainly involves Cα-hydroxyl oxidation, alkyl-aryl cleavage, and demethoxylation (Breen and Singleton 1999).When small molecular mediators such as hydroxyl-benzotriazole are present, laccases are also able to degrade nonphenolic lignin structures via one electron oxidation (Call and Muche 1997).

Similar to LiP and MnP, white rot fungi harbor several laccase (*lcc*) genes that encode multiple isoforms. Copper, nitrogen, and certain aromatic compounds have been tested (i.e., 2, 5-xylidine and 1-hydroxybenzotriazole) and reported to activate *lcc* transcriptions (Colllins and Dobson 1997). Mansur et al. (1998) discovered that *lcc1* and *lcc2* were induced by veratryl alcohol at different stages of growth while *lcc3* were noninduced by veratryl alcohol.

3.3.4 VersatileP eroxidase

Versatile peroxidases (VPs), which have been found and characterized in various *Pleurotus* and *Bjerkandera* species, are regarded as a hybrid peroxidase because they have both LiP and MnP activity (Camarero et al. 1999; Mester and Field 1998). Correspondingly, the structures of VPs share both MnP and LiP substrate interaction sites, such as Mn^{2+} binding sites for Mn^{2+} oxidation exhibited by MnPs (Camarero et al. 1999) and invariant tryptophan residues responsible for direct one-electron oxidation of nonphenolics via LiPs (Perez-Boada et al. 2005). The molecular weight of VPs, around 40–50 kDa, is also close to that of LiPs and MnPs. Thus, similar to MnP and LiP, VPs are capable of degrading phenolic and nonphenolic compounds (Martinez et al. 1996; Mester and Field 1998). However, VPs do not functionally require H_2O_2, which is different from LiPs and MnPs. Instead, these peroxidases generate H_2O_2 by oxidation of hydroquinone in the presence of Mn^{2+} (Gómez-Toribio et al. 2001). VPs could also possess oxidative abilities that neither LiPs nor MnPs have. For example, substituted phenol was efficiently oxidized by VPs isolated from *Pleurotus eryngii* but not by peroxidases from *P.c hrysosporium* (Martinez et al. 1996).

3.3.5 Peroxide-ProducingE nzymes

Extracellular H_2O_2 is required for activating LiP and MnP for subsequent lignnolysis. H_2O_2 generating enzymes, such as glyoxal oxidase (GLOX), aryl alcohol oxidase (AAO), and glucose-1-oxidase (GOX), have been discovered in many white rot fungi for supplying extracellular H_2O_2 (Cohen et al. 2002). For example, GLOX accepts electrons from aldehyde-type substrates by coupling reduction of O_2 to H_2O_2 (Kirk and Cullen 1998). Extracellular metabolites (e.g., glyoxal, methylglyoxal) produced by some white rot fungi, or lignin degraded fragments such as glycolaldehyde, can serve as the substrates for GLOX (Hammel et al. 1994; Kersten 1990). Although extracellular oxidases are the primary sources of H_2O_2, intracellular sugar oxidase found in a few fungi has been proposed to be involved in H_2O_2 supply (Kirk and Farrell 1987).

3.4 FungalPr etreatmentPr ocess

Solid-state fungal pretreatment involves degradation of the feedstock that occurs without free water (Gowthamana et al. 2001). Processing parameters, such as moisture, temperature, and aeration, are crucial for lignin degradation. In general, a moist environment favors fungal growth and activity as nutrients can be readily diffusible and accessible by the fungi (Reid 1989). Although optimal moisture content varies with strains and substrates, prior studies suggested that fungi can grow well and substantially degrade lignin with an initial moisture content of 60–85 % (Wan and Li 2012). Some studies have reported that the initial moisture content for fungal pretreatment can be as low as 50–60 %, depending on the feedstock and nutrient amendment. Too high a moisture may favor formation of fungal mycelia while inhibiting the delignification process (Zadrazil and Brunnert 1981). However, too low a moisture content tends to retard fungal growth. Although fungal metabolism increases moisture due to production of water, it does not help maintain moisture levels due to significant moisture loss under natural ventilation conditions. For example, Cui et al. (2012) reported that the moisture content gradually decreased from 75 to 5 % during 90 days of pretreatment of corn stover with *C. subvermispora*. After 50 days, the moisture content was below 45 %, where no further improvement on sugar yield was observed as the fungal growth and metabolism almost ceased at such a low moisture content. Due to lack of free water and low conductivity of solid particles, heat generated by fungal metabolic activity generally causes temperature gradients in the reactors. Accumulated heat should be dissipated via a way that ensures suitable temperatures for ascomycetes, around 39 °C, and for white rot basidiomycetes, between 20 and 30 °C (Reid 1989). Aeration is necessary to dissipate heat, but more importantly, to provide uniform air/oxygen diffusion throughout the substrate. Oxygen enrichment could increase the delignification rate but may not affect delignification selectivity (Hatakka 1983; Reid 1989). In addition, the aeration rate also affects fungal performance. In a study of *P. chrysosporium* treatment of aspen chips for biopulping, the median aeration rate of three test levels ($0.001, 0.022, 0.1$ v v^{-1} min^{-1}) was enough to achieve good fungal growth and degradation (Messner et al. 1998). Thus, aeration needs to be controlled to ensure effectiveness of biological pretreatment. Nutrient supplements, such as nitrogen sources, Mn^{2+}, and aromatic compounds, also play important roles in fungal degradation. Similar to physical culture conditions, nutrients affect fungal growth and differentially regulate expression of white rot fungi genes encoding ligninolytic enzymes. However, the response of white rot fungi to nutrient addition varies between species and strains due to regulatory elements in promoter regions of genes (Collins and Dobsen 1997; Have and Teunissen 2001; Janse et al. 1998).

Taking into consideration the effects of processing parameters, various bioreactors, including tray reactors, packed-beds, rotating drums, and stirred bioreactors, have been developed for solid-state fermentation (Mitchell et al. 2006). Mycelial fungi act on the substrate by binding the solid particles with the interparticle hyphal bridges across the substrates. Tray and packed-bed reactors which are operated statically are used for mycelia fungi as mixing can disrupt hyphae between particles (Fanaei and Vaziri 2009; Mitchell et al. 2000). The tray reactors are simple but have a limited loading capacity due to a low production volume. Packed-bed reactors, with a larger loading capacity, are widely used for solid-state fermentation, but have a major disadvantage of temperature gradients. Convective heat transfer, instead of water jacked such as forced humid air, is used for tradi- tional packed-beds for the purpose of removing the heat while preventing the substrate bed from drying out. The main problems with this design are axial tem- perature gradients and evaporative losses (Mitchell et al. 2000). One of the modi- fied packed-bed bioreactors, the zymotis packed-bed, which is equipped with internal heat transfer plates, was developed to overcome these problems (Mitchell et al. 2006). However, this type of reactor also increases operational costs and diffi- culties due to the presence of the heat transfer plates. For fungal pretreatment, tubu- lar reactors such as PVC tubes and silos were used. As some white rot fungi can outcompete or co-exist with indigenous microorganisms, relatively rough and open reactors, such as chip piles were tested for scale-up of fungal pretreatment (Scott et al. 1998). Although pile reactors had a lower cost than packed bed reactors, dif- ficulty in managing heat transfer in large-scale pile reactors (up to 40 tons) led to a lower quality of pretreated feedstock due to degradation caused by undesirable bac- teria and fungi. Therefore, more practical bioreactors that ease both operation and heat transfer are under development for solid-state fungal pretreatment.

3.5 EffectofFungalPr etreatmentonL ignocellulosic Biomass

White rot fungi enzymatically "combust" the lignin in biomass feedstocks while consuming carbohydrates via hydrolysis of self-secreted cellulolytic and/or hemi- cellulolytic enzymes. Analysis of white rot decayed feedstocks shows substantially altered physiochemical characteristics of the biomass. Ideally, most of the cellu- lose remains after delignification and its digestibility is improved significantly. However, experimental results indicate that improvement of cellulose digestibility is highly dependent on degradation patterns of the fungi. The remaining cellulose is more susceptible to subsequent enzymatic hydrolysis when fungi preferentially degrade lignin, while it is less digestible when fungi simultaneously degrade lignin and cellulose.

Table 3.1 summarizes the effect of fungal pretreatment on digestibility of lig- nocellulosic biomass. As a widely studied but nonselective white rot fungus, *P. chrysosporium* degraded both lignin and holocellulose (cellulose and hemicellu- lose) to a large extent, but resulted in little or no improvement of cellulose digest- ibility. For example, lignin and holocellulose loss in aspen wood increased with pretreatment time, reaching 42 and 17 %, respectively, after 28 days of *P. chrys- osporium* pretreatment (Sawada et al. 1995). Extension of the pretreatment time to 100 days led to 8 % more lignin removal but 33 % more holocellulose

Table 3.1 The effect of solid-state biological pretreatment on enzymatic hydrolysis and ethanol production

Fungus	Substrate	Sugar/ethanol yield[a]	Reference
Phaerochaete chrysosporium	Cotton stalk	Reduced glucose yield	Shi et al. (2009)
Phaerochaete chrysosporium	Corn stover	No improvement on glucose yield	Keller et al. (2003)
Phaerochaete chrysosporium	Beech wood	9.5 % total sugar yield[b]	Sawada et al. (1995)
Phaerochaete chrysosporium	Corn fiber	Reduced sugar yield or no significant improvement	Shrestha et al. (2008)
Pleurotus ostreatus	Rice straw	33 % glucose yield	Taniguchi et al. (2005)
Pleurotus ostreatus	Rice hull	38.9 % glucose yield	Yu et al. (2009b)
Pleurotus ostreatus, Pycnoporus cinnabarinus 115	Wheat straw	27–28 % glucose yield[c]	Hatakka (1983)
Euc-1	Wheat straw	22.5 % total sugar yield[b]	Dias et al. (2010)
Cyathus stercoeus	Corn stover	36 % glucose yield	Keller et al. (2003)
Irpex lactues	Corn stover	66.4 % total sugar yield[d]	Xu et al. (2010)
Ceriporiopsis subvermispora	Corn stover	56–66 % glucose yield and 57.80 % ethanol yield	Wan and Li (2010a, b)
Pheblia tremellosus	Aspen wood	11.6 % glucose yield[b]	Mes-Hartree et al. (1987)
Polyporus giganteus	Aspen wood	55.2 % glucose yield	Kirk and Moore (1972)
Stereum hirsutum	Japanese red pine	13.56 % glucose yield[b]	Lee et al. (2007)
Echinodontium taxodii 2538	Chinese willow, China-fir	5–35 % glucose yield	Yu et al. (2009a)
Echinodontium taxodii 2538, *Coriolus versicolor*	Bamboo culm	37 % total sugar yield[e]	Zhang et al. (2007a, b)

Reprinted from Biotechnology Advances, Caixia Wan, Yebo Li, Fungal pretreatment of lignocellulosic biomass, 1, Copyright 2012, with permission from Elsevier

[a] % of theoretical yield of glucan in the original material, unless stated otherwise; [b] % of dry mass of the treated material; [c] % of dry mass of the original material; [d] % of theoretical yield of holocellulose in the original material; [e] % of theoretical yield of holocellulose in the treated material

degradation. The resulting sugar yields increased during the first 28 days and then kept decreasing due to degradation of large amounts of holocellulose during prolonged pretreatment periods. The results indicated that pretreatment with *P. chrysosporium* was not sufficient to enhance enzymatic hydrolysis of aspen wood meal. Similar findings were reported for cotton stalks (Shi et al. 2009), corn fiber (Shrestha et al. 2008), and corn stover (Keller et al. 2003). Reduced sugar yields after *P. chrysosporium* pretreatment were also observed with those biomass feedstocks. Shi et al. (2009) reported that for a 14-day pretreatment, *P. chrysosporium* degraded 36 % lignin but at the expense of 40 % cellulose loss. The authors also washed treated material using hot water with an attempt to remove possible inhibitors to enzymatic hydrolysis which may have resulted from fungal pretreatment. However, the glucose yield was still close to that of non-treated. Therefore, nonselective fungi may not be suitable for improving enzymatic hydrolysis due to their vigorous cellulose degradation. In contrast, selective lignin degrading fungi preferentially degrade lignin while preserving most cellulose and in turn result in significantly increased digestibility. For example, lignin loss of corn stover was 36 % during 35 days of pretreatment with *C. subvermispora*, a typical selective fungus, while cellulose loss was less than 5 % throughout fungal pretreatment (Wan and Li 2010a, b). At an enzyme loading of 10 FPU/g solid for enzymatic hydrolysis, the glucose yield of fungal-pretreated corn stover was as high as 67 %, which was aboutthre etime stha tofnontre ated.

The ability of white rot fungi to delignify lignocellulosic biomass varies among genera and species as well as in various fungal-substrate combinations (Akin et al. 1995; Anderson and Akin 2008). Herbaceous plants contain guaiacyl and syringlyl lignin and significant amounts of *P*-hydroxyphenyl lignin (10–20 %) (Boerjan et al. 2003). High contents of ester-linked *p*-coumeric and ferulic units were also found in nonlignified cell walls, which are believed to partially contribute to grass recalcitrance. However, in delignification of Bermudagrass, *C. subvermispora* and *Cythus stercoreus* were capable of attacking both ester- and ether-linked phenolic acids from unlignified cell walls and also showed increased degradation of guaiacyl lignin over syringyl lignin (Akin et al. 1995). In contrast, in the degradation of woody biomass by some white rot fungi, such as *C. subvermispora*, the results indicated that syringyl (S)-rich lignin was degraded more rapidly than other lignin subunits (e.g., hydroxyphenyl (H), guaiacyl (G)) (Camarero et al. 1999; Eriksson et al. 1990). In other studies, hardwoods, which consist of about equal amounts of guaiacyl and syringyl lignin, appeared to be more susceptible to fungal degradation than softwoods that contain a high amount of guaiacyl lignin (90 %). For example, 50–55 % of the polysaccharides in aspen wood were enzymatically converted into simple sugars after pretreatment with *Polyporus giganteus, Polyporus berkeleyi,* or *Polyporus resinosus* for 63–99 days (Kirk and Moore 1972). Chinese willow, after pretreatment with *E. taxodii* 2538 for 120 days, also resulted in about 37 % sugar yields at a cellulase loading of 20 FPU/g solid for enzymatic hydrolysis. In contrast, even with a pretreatment time of more than 8–12 weeks, cellulose digestibility of fungal-treated softwood, such as China fir (Yu et al. 2009a) and Japanese red pine (Lee et al. 2007), was less than 20 %. Studies of agricultural residues have shown that improvement in digestibility due to fungal pretreatment varies with different feedstocks. Wan and Li (2011a) compared 18-day fungal degradation on different types of feedstock and found that among those tested, *C. subvermispora* was especially effective for corn stover, switchgrass, and hardwood, with lignin degradation of 28, 26, and 15 %, respectively, while the fungus was not effective for two other agricultural residues, wheat straw and soybean straw. Similarly, *C. subvermispora* was also less effective for delignification of rice straw, likely due to high contents of *p*-coumeric units in lignin of rice straw (Taniguchi et al. 2005). In contrast, *P. ostreatus* appeared to be more effective on straw materials than other fungi and has been widely tested for biological pretreatment of these kinds of feedstocks (Hatakka 1983; Taniguchi et al. 2005; Yu et al. 2009b). These variations in fungal performance on different feedstocks suggest that specific substrate-fungal interactions occur. Various factors, such as lignin compositions, covalent bonds linking lignin and hemicelluloses, and plant extractives, may restrict the degradation of plant cell walls by the fungi (Grabber 2005).

3.6 SynergismB etweenF ungalandPh ysicochemical Pretreatment

Biological pretreatment with white rot fungi has been traditionally applied in the biopulping industry to convert wood chips into paper pulp. Its benefits include saving energy and/or reducing reaction severity in the subsequent mechanical/chemical pulping. Similarly, fungal pretreatment has been studied in combination with mild mechanical or physical/chemical pretreatments in order to synergistically improve the digestibility of lignocellulosic biomass while overcoming disadvantages associated with either pretreatment method. Specifically, fungal pretreatment time can be shortened while delignification efficiency can be increased. In one study, a relatively short fungal incubation time (1–2 weeks) only modified the lignin structure rather than depolymerizing it, but substantially improved the subsequent chemical pulping process (Messner et al. 1998). A similar observation was reported with fungal pretreatment followed by mild alkaline or acid pretreatment of lignocellulosic biomass. Yu et al. (2010) reported that pretreatment of corn stalks with *I. lacteus* for only 15 days shortened either pretreatment time or pretreatment temperature of the subsequent alkaline pretreatment, which was conducted at 1.5 % alkaline loading. For pretreatment of water hyacinth, fungal pretreatment for 15 days improved its sugar yields by 1–2 times over dilute sulfuric acid pretreatment alone (Ma et al. 2010). Similarly, combined *C. subvermispora* pretreatment and ethanolysis improved ethanol yield of woody biomass by 1–2 times over a single pretreatment (e.g., Japanese cedar, beech wood) (Baba et al.2011 ;Itoh e ta l.2003).

Fungal pretreatment following non-biological pretreatment also been shown to improve the efficiency as the first step of this combined pretreatment at mild conditions generally reduces biomass recalcitrance to fungal pretreatment. As a result, the time of fungal pretreatment can be shortened. Yu et al. (2009b) reported that the sugar yield resulting from 18-day fungal pretreatment succeeding ultrasonic pretreatment (250 w, 40 kHz, 30 min) was comparable to that from 42-day fungal pretreatment alone. It is also interesting to note that for some premodified biomass, fungal pretreatment can become effective for feedstocks that were originally resistant to the white rot fungi. The study of Wan and Li (2011b) showed that hot water pretreatment at 170 °C for 3 min facilitated fungal pretreatment of soybean straw while single fungal pretreatment was not effective for this feedstock. The glucose yields resulting from combined pretreatment were improved by 30 % over a sole pretreatment. Analysis of pretreated soybean straw showed that the cell wall structure was modified rather than degraded by hot water pretreatment at the conditions tested. On the other hand, combination of hot water and fungal pretreatment did not improve corn stover which can be readily degraded by *C. subvermispora*, probably because fungal pretreatment was effective without other pretreatments and thus masked possible synergism from adding a hot water pretreatment. These, as well as other studies, have suggested that biomass feedstocks most suitable for fungal pretreatment as the second step of combined pretreatment should be resistant to fungal treatment to some extent. In addition, regardless of pretreatment order, the mildest physicochemical pretreatment that produces the sufficient modification is suggested.

3.7 Summary

Biological pretreatment with white rot fungi under solid-state conditions reduces biomass recalcitrance through unique ligninolytic enzymes. In general, due to a slow growth rate of the fungi, a few weeks to months are required to achieve significant delignification, which becomes the main drawback of this technology. However, benefits of this pretreatment are obvious, as it does not use chemicals and also minimizes downstream waste treatment. Such "green" pretreatment has been of particular interest for on-farm scale applications because on-farm in-storage pretreatment can generally incorporate the long pretreatment times needed to maintain a year-round supply of pretreated feedstock for biorefineries. Ensilage, which harnesses indigenous lactic acid bacteria for degrading cell walls under solid-state conditions, is traditionally used for feed upgrading on farms and recently has been studied for the pretreatment of lignocellulosic biomass for biofuels. Fungal pretreatment by white rot fungi has advantages over ensilage as it provides a delignified but cellulose-rich biomass residue. In order to improve fungal performance, proper processing conditions should be optimized and incorporated into the reactor design. In addition to being used as a single pretreatment, synergism arising from pretreatment combinations suggests fungal pretreatment could be a good supplement to thermochemical pretreatment by overcoming shortcomings of both methods. The benefits include shortened fungal pretreatment time and reduced severity of thermochemical pretreatments, while increasing digestibility of pretreated biomass. On the other hand, fungal pretreatment alone can be applied to on-farm wet storage by taking advantage of long pretreatment time while improving cellulose digestibility. In conclusion, solid-state fungal-pretreatment, with or without combination of mild thermo-chemical pretreatment, has the potential to be an important alternative for biofuel production.

References

Akin DE, Rigsby LL, Sethuraman A, Morrison WH, Gamble GR, Eriksson KEL (1995) Alterations in structure, chemistry, and biodegradability of grass lignocellulose treated with the white-rot fungi *Ceriporiopsis-Subvermispora* and *Cyathus-Stercoreus*. Appl Environ Microbiol61:1591–1598

Anderson W, Akin D (2008) Structural and chemical properties of grass lignocelluloses related to conversion for biofuels. J Ind Microbiol Biotechnol 35:355–366

Baba Y, Tanabe T, Shirai N, Watanabe T, Honda Y, Watanabe T (2011) Pretreatment of Japanese cedar wood by white rot fungi and ethanolysis for bioethanol production. Biomass Bioenerg 35:320–324

Blanchette RA (1995) Degradation of the lignocellulose complex in wood. Can J Bot 73:S 999–S101

Blanchette RA, Krueger EW, Haight JE, Akhtar M, Akin DE (1997) Cell wall alterations in loblolly pine wood decayed by the white-rot fungus, *Ceriporiopsis subvermispora*. J Biotechnol53:203 –213

Boerjan W,R alphJ ,B aucherM (2003)Li gninbi osynthesis. AnnuR evP lantB iol54: 519–546

Bogan BW, Schoenike B, Lamar RT, Cullen D (1996) Expression of *lip* genes during growth in soil and oxidation of anthracene by *Phanerochaete chrysosporium*. Appl Environ Microbiol 62:3697–3703

Bollag JM, Leonowicz A (1984) Comparative studies of extracellular fungal laccases. Appl Environ Microbiol 48:849–854

Breen A, Singleton FL (1999) Fungi in lignocellulose breakdown and biopulping. Curr Opin Biotechnol10:252 –258

Call HP, Mücke I (1997) History, overview, and applications of mediated lignolytic systems, especially laccase-mediator-systems (lignozyme-process). J Biotechnol 53:163–202

Camarero S, Sarkar S, Ruiz-Dueñas FJ, Martínez MJ, Martínez AT (1999) Description of a versatile peroxidase involved in the natural degradation of lignin that has both manganese peroxidase and lignin peroxidase substrate interaction sites. J Biol Chem 274:10324–10330. doi:10.1074/jbc.274.15.10324

Campbell MM, Sederoff RR (1996) Variation in lignin content and composition—Mechanism of control and implications for the genetic improvement of plants. Plant Physiol 110:3–13

Chen CL, Chang H, Kirk TK (1983) Carboxylic-acids produced through oxidative cleavage of aromatic rings during degradation of lignin in spruce wood by *Phanerochaete-chrysosporium*.J WoodC hem Technol3: 35–57

Cohen R, Persky L, Hadar Y (2002) Biotechnological applications and potential of wood-degradation mushrooms of the genus *Pleurotus*. Appl Microbiol Biotechnol 58:582–594. doi:10.1007/s00253-002-0930-y

Collins PJ, Dobson ADW (1997) Regulation of laccase gene transcription in *Trametes versicolor*. ApplEn vironM icrobiol63: 3444–3450

Cui Z, Shi J, Wan C, Li Y (2012) Comparison of alkaline- and fungi-assisted wet-storage of corn stover.B ioresour Technol109: 98–104

Dhawale SS (1993) Is an activator protein-2-like transcription factor involved in regulating gene-expression during nitrogen limitation in fungi. Appl Environ Microbiol 59:2335–2338

Dias AA, Freitas GS, Marques GSM, Sampaio A, Fraga IS, Rodrigues MAM, Evtuguin DV, Bezerra RMF (2010) Enzymatic saccharification of biologically pre-treated wheat straw with white-rot fungi. Bioresour Technol101: 6045–6050

Digman MF, Shinners KJ, Casler MD, Dien BS, Hatfield RD, Jung HJG, Muck RE, Weimer PJ (2010) Optimizing on-farm pretreatment of perennial grasses for fuel ethanol production. Bioresour Technol101: 5305–5314

Dosoretz CG, Ward G, Hadar Y (2004) Lignin peroxidase. In: Pandey A (ed) Concise encyclopedia of bioresource technology. The Haworth Press, New York

Doyle WA, Boldig W, Veitch NC, Piontek K, Smith AT (1998) Two substrate interaction sites in lignin peroxidases revealed by site-directed mutagenesis. Biochemistry 37:15097–15105

Enoki M, Watanabe T, Nakagame S, Koller K, Messner K, Honda Y, Kuwahara M (1999) Extracellular lipid peroxidation of selective white-rot fungus, *Ceriporiopsis subvermispora*. FEMSM icrobiolL ett180: 205–211

Eriksson KEL, Johnsrud SC, Vallander L (1983) Degradation of lignin and lignin model compounds by various mutants of the white-rot fungus *Sporotrichum pulverulentum*. Arch Microbiol135:161–168

Eriksson KEL, Blanchette RA, Ander P (1990) Microbial and enzymatic degradation of wood and wood components. Springer, New York

Fanaei MA, Vaziri BM (2009) Modeling of temperature gradients in packed-bed solid-state bioreactors. ChemEn gPr ocess48: 446–451

Farrell RL, Murtagh KE, Tien M, Mozuch MD, Kirk TK (1989) Physical and enzymatic properties of ligninperoxidase isoenzymes from *Phanerochaete chrysosporium*. Enzyme Microb Technol11: 322–328

Galkin S, Vares T, Kalsi M, Hatakka A (1998) Production of organic acids by different white-rot fungi as detected using capillary zone electrophoresis. Biotechnol Tech 12:267–271

Gaskell J, Stewart P, Kersten PJ, Cover SF, Reiser J, Cullen D (1994) Establishment of genetic linage by allel-specific peroxidase gene family of *Phanerochaete chrysoporium*. Biotechnology12 :1372–1375

Gianfreda L, Xu f, Bollag JM (1999) Laccases: a useful group of oxidoreductive enzymes. Bioremediation J 3:1–25

Gilardi G, Harvey PJ, Cass AEG, Palmer JM (1990) Radical intermediates in veratryl alcohol oxidation by ligninase-NMRe vidence.B iochimB iophys Acta1 041:129–132

Gold MH, Youngs HL, Sollewijn Gelpke MD (2000) Manganese peroxidase. In: Sigel A, Sigen H (eds) Metal ions biological systems. Marcel Dekker, New York

Gómez-Toribio V, Martínez A, Martínez MJ, Guillén F (2001) Oxidation of hydroquinones by versatile ligninolytic peroxidase from *Pleurotus eryngii*-H_2O_2 generation and the influence of Mn^{2+}. Eur J Biochem 268: 4787–4793

Gowthamana MK, Krishnab C, Young MM (2001) Fungal solid state fermentation—an overview. ApplM ycolB iotechnol1: 305–352

Grabber JH (2005) How do lignin composition, structure, and cross-linking affect degradability? A review of cell wall model studies. Crop Sci 45:820–831

Green F, Highley TL (1997) Mechanism of brown-rot decay: paradigm or paradox. Int Biodeter Biodegr39: 113–124

Hammel KE, Cullen D (2008) Role of fungal peroxidases in biological ligninolysis. Curr Opin Plant Biol 11:349–355

Hammel KE, Kalyanaraman B, KirkTK (1986) Substrate free radicals are intermediates in ligninase catalysis. Proc Natl AcadSc iU SA83: 3708–3712

Hammel KE, Jensen KA, Mozuch MD, Landucci LL, Tien M, Pease EA (1993) Ligninolysis by a purified lignin peroxidase. J Biol Chem 268:12274–12281

Hammel KE, Mozuch MD, Jensen KA, Kersten PJ (1994) H_2O_2 recycling during oxidation of the arylglycerol β-aryl ether lignin structure by lignin peroxidase and glyoxal oxidase. Biochemistry33:13349–13354

Hatakka AI (1983) Pretreatment of wheat straw by white-rot fungi for enzymic saccharification of cellulose. Appl Environ Microbiol 18:350–357

Hatakka A (1994) Lignin-modifying enzymes from selected white-rot fungi: production and role from in lignin degradation. FEMS Microbiol Rev 13:125–135

Have R, Teunissen PJM (2001) Oxidative mechanisms involved in lignin degradation by white-rot fungi. Chem Rev 101:3397–3413

Isroi RM, Syamsiah S, Niklasson C, Cahyanto MN, Lundquist K, Taherzadeh MJ (2011) Biological pretreatment of lignocelluloses with white-rot fungi and its applications: a review. Bioresources6:5224–5259

Itoh H, Wada M, Honda Y, Kuwahara M, Watanabe T (2003) Bioorganosolve pretreatments for simultaneous saccharification and fermentation of beech wood by ethanolysis and white rot fungi. J Biotechnol 103:273–280

Janse BJH, Gaskell J, Akhtar M, Cullen D (1998) Expression of *Phanerochaete chrysosporium* genes encoding lignin peroxidases, manganese peroxidases, and glyoxal oxidase in wood. ApplEn vironM icrobiol64: 3536–3538

Kapich AN, Jensen KA, Hammel KE (1999) Peroxyl radicals are potential agents of lignin bio-degradation.FEB SLe tt461: 115–119

Keller F, Hamilton J, Nguyen Q (2003) Microbial pretreatment of biomass. Appl Biochem Biotechnol105:27 –41

Kersten PJ (1990) Glyoxal oxidase of *Phanerochaete chrysosporium*: its characterization and activation by lignin peroxidase. Proc Natl Acad Sci USA 87:2936–2940

Kirk TK, Cullen D (1998). Enzymology and molecular genetics of wood degradation by white-rot fungi. In: Young RA, Akhtar M (eds) Environmental friendly technologies for the pulp and paper industry. Wiley, New York

Kirk TK, Farrell RL (1987) Enzymatic combustion—the microbial-degradation of lignin. Annu Rev Microbiol 41:465–505

Kirk TK, Highley TL (1973) Quantitative changes in structural components of conifer woods during decay by white-rot and brown-rot fungi. Phytopathology 63:1338–1342

Kirk T, Moore W (1972) Removing lignin from wood with white-rot fungi and digestibility of resulting wood. Wood Fiber Sci 4:72–79

Lee JW, Gwak KS, Park JY, Park MJ, Choi DH, Kwon M, Choi IG (2007) Biological pretreatment of softwood *Pinusde nsifbr a* by three white rot fungi. J Microbiol 45:485–491

Ma F, Yang N, Xu C, Yu H, Wu J, Zhang X (2010) Combination of biological pretreatment with mild acid pretreatment for enzymatic hydrolysis and ethanol production from water hyacinth. Bioresour Technol101: 9600–9604

Mansur M, Suarez T, Gonzalez AE (1998) Differential gene expression in the laccase gene family from basidiomycete I-62 (CECT20197) . ArchB iochemB iophys64: 771–774

Martínez MJ, Ruiz-Dueñas FJ, Guillén F, Martínez AT (1996) Purification and catalytic properties of two manganese-peroxidase isoenzymes from *Pleurotus eryngii*. Eur J Biochem 237:424–432

Mes-Hartree M, Yu EKC, Reid ID, Saddler JN (1987) Suitability of aspen wood biologically delignified with Pheblia tremellosus for fermentation to ethanol or butanediol. Appl Microbiol Biotechnol26:120–125

Messner K, Koller K, Wall MB, Akhtar M, Scott GM (1998) Fungal treatment or wood chips for chemical pulping. In: Young RA, Akhtar M (eds) Environmental friendly technologies for the pulp and paper industry. Wiley, New York

Mester T, Field JA (1998) Field characterization of a novel manganese peroxidase-lignin peroxidase hybrid isozyme produced by *Bjerkandera* species strain BOS55 in the absence of manganese. J Biol Chem 273:15412–15417

Millis CD, Cai DY, Stankovich MT, Tien M (1989) Oxidation-reduction potentials and ionization states of extracellular peroxidases from the lignin-degrading fungus *Phanerochaete chrysosporium*. Biochemistry 28:8484–8489

Mitchell DA, Krieger N, Stuart DM, Pandey A (2000) New developments in solid-state fermentation: II. Rational approaches to the design, operation and scale-up of bioreactors. Process Biochem35:1211–1225

Mitchell DA, Krieger N, Berovic M (2006) Solid-state fermentation bioreactors: fundamentals of design and operation. Springer-Verlag, New York

Mosier N, Wyman CE, Dale B, Elander R, Lee YY, Holtzapple M, Ladisch M (2005) Features of promising technologies for pretreatment of lignocellulosic biomass. Bioresour Technol 96:673–686

Pérez-Boada M, Ruiz-Dueñas FJ, Pogni R, Basosi R, Choinowski T, Martínez MJ, Piontek K, Martínez AT (2005) Versatile peroxidase oxidation of high redox potential aromatic compounds: site-directed mutagenesis, spectroscopic and crystallographic investigations of three long-range electron transfer pathways. Mol Biol 354:385–402

Rajakumar S, Gaskell J, Cullen D, Lobos S, Karahanian E, Vicuna R (1996) *Lip*-like genes in *Phanerochaete sordida*, and *Ceriporiopsis subvermispora*, white rot fungi with no detectable lignin peroxidase activity. Appl Environ Microbiol 62:2660–2663

Reid ID (1989) Solid-state fermentations for biological delignification. Enzyme Microb Technol 11:786–803

Sawada T, Nakamura Y, Kobayashi F, Kuwahara M, Watanabe T (1995) Effects of fungal pretreatment and steam explosion pretreatment on enzymatic saccharification of plant biomass. BiotechnolB ioeng48: 719–724

Scott GM, Akhtar M, Lentz MJ, Swaney RE (1998) Engineering, scale-up, and economic aspects of fungal pretreatment of wood chips. In: Young RA, Akhtar M (eds) Environmental friendly technologies for the pulp and paper industry. Wiley, New York

Shi J, Sharma-Shivappa RR, Chinn M, Howell N (2009) Effect of microbial pretreatment on enzymatic hydrolysis and fermentation of cotton stalks for ethanol production. Biomass Bioenerg33:88–9 6

Shinners KJ, Binversie BN, Muck RE, Weimer P (2007) Comparison of wet and dry corn stover harvest and storage. Biomass Bioenerg31: 211–221

Shrestha P, Rasmussen M, Khanal SK, Pometto AL, van Leeuwen J (2008) Solid-substrate fermentation of corn fiber by *Phanerochaete chrysosporium* and subsequent fermentation of hydrolysate into ethanol. J Agric Food Chem 56:3918–3924

Srebotnik E, Messner K, Foisner R (1988) Penetrability of white rot-degraded pine wood by the lignin peroxidase of *Phanerochaete-chrysosporium*. ApplE nvironM icrobiol54: 2608–2614

Taniguchi M, Suzuki H, Watanabe D, Sakai K, Hoshino K, Tanaka T (2005) Evaluation of pretreatment with *Pleurotus ostreatus* for enzymatic hydrolysis of rice straw. J Biosci Bioeng 100:637–643

Varm A, Kolli BK, Paul J, Saxena S, König H (1994) Lignocellulose degradation by microorganisms from termite hills and termite guts: A survey on the present state of art. FEMS Microbiol Rev 15:9–28

Wan C (2011) Microbial pretreatment of lignocellulosic biomass with *Ceriporiopsis Subvermispora* for enzymatic hydrolysis and ethanol production. Dissertation, The Ohio State University

Wan C, Li Y (2010a) Microbial pretreatment of corn stover with *Ceriporiopsis subvermispora* for enzymatic hydrolysis and ethanol production. Bioresour Technol101: 6398–6403

Wan C, Li Y (2010b) Microbial pretreatment of corn stover with *Ceriporiopsis subvermispora* for enzymatic hydrolysis and ethanol production. Bioresour Technol101: 6398–6403

Wan C, Li Y (2011a) Effectiveness of microbial pretreatment by *Ceriporiopsis subvermispora* on different biomass feedstocks.B ioresour Technol102: 7507–7512

Wan C, Li Y (2011b) Effect of hot water extraction and liquid hot water pretreatment on the fungald egradationo fb iomassf eedstocks.B ioresour Technol1 02:9788–9793

Wan C, Li Y (2012) Fungal pretreatment of lignocellulosic biomass. Biotechnol Adv. doi:10.1016/j.biotechadv.2012.03.003

Watanabe H, Tokuda G (2001) Animal cellulases. Cell Mol Life Sci 58:1167–1178. doi:10.1007/PL00000931

Watanabe T, Katayama S, Enoki M, Honda YH, Kuwahara M (2000) Formation of acyl radical in lipid peroxidation of linoleic acid by manganese-dependent peroxidase from *Ceriporiopsis subvermispora* and *Bjerkanderaadus ta*. Eur J Biochem 267:4222–4231

Wong D (2009) Structural and action mechanism of ligninolytic enzymes. Appl Biochem Biotech 157:174–209

Xu C, Ma F, Zhang X, Chen S (2010) Biological pretreatment of corn stover by *Irpex lacteus* for enzymatic hydrolysis. J Agric Food Chem 58:10893–10898

Yu H, Guo G, Zhang X, Yan K, Xu C (2009a) The effect of biological pretreatment with the selective white-rot fungus *Echinodontium taxodii* on enzymatic hydrolysis of softwoods and hardwoods.B ioresour Technol100: 5170–5175

Yu J, Zhang J, He J, Liu Z,Y u Z (2009b) Combinations of mild physical or chemical pretreatment with biological pretreatment for enzymatic hydrolysis of rice hull. Bioresour Technol 100:903–908

Yu H, Du W, Zhang J, Ma F, Zhang X, Zhong W (2010) Fungal treatment of corn stalks enhances the delignification and xylan loss during mild alkaline pretreatment and enzymatic digestibility of glucan. Bioresour Technol 101:6728–6734

Zadrazil F, Brunnert H (1981) Investigation of physical parameters important for the solid-state fermentation of straw by white rot fungi. Eur J Appl Microbiol Biotechnol 11:183–188

Zhang X, Xu C, Wang H (2007a) Pretreatment of bamboo residues with *Coriolus versicolor* for enzymatich ydrolysis.J B iosciB ioeng1 04:149–151

Zhang X, Yu H, Huang H, Liu Y (2007b) Evaluation of biological pretreatment with white rot fungi for the enzymatic hydrolysis of bamboo culms. Int BiodeterB iodegr60: 159–164

Chapter 4
HydrothermalPr etreatment
of Lignocellulosic Biomass

IwonaC ybulska, GrzegorzB rudeckiand HanwuL ei

Abstract As global energy demands grow and as the environmental and economic issues of fossil fuel use arise, lignocellulosic biomass is starting to attract increased attention as a potential source of energy and chemicals. Being an abundant, accessible, and cost-effective feedstock for a wide variety of products, ranging from transportation fuels to pharmaceuticals, lignocellulose shows great promise for the future. However, in order to utilize its potential, an efficient pretreatment method has to be applied. Hydrothermal pretreatment is one of the most promising and environmentally friendly biomass pretreatment methods available to make the lignocellulosic biomass vulnerable to enzymatic breakdown. This chapter describes the principle of the hydrothermal pretreatments, as well as influence of temperature and time on the effectiveness of the pretreatment and the kinetic models of the process. Various configurations of systems employing hydrothermal pretreatments have also been presented (with examples of process conditions), including hot water, steam explosion, catalyzed hydrothermal treatment, and combination with other methods.

Keywords Hydrothermal pretreatment • Hot water • Steam explosion • Lignocellulosic biomass

I.C ybulska
ChemicalEngi neeringPr ogram,M asdarI nstituteof S ciencea nd Technology,
AbuD habi,U nited ArabEm irates

G. Brudecki
Departmentof Agriculturala ndB iosystemsEngi neering,So uthD akotaS tateU niversity,
Brookings,SD 57007 ,U SA

H. Lei(✉)
Department of Biological Systems Engineering, Washington State University,
Richland, WA 99354, USA
e-mail:hle i@wsu.edu

T. Gu(e d.), *Green Biomass Pretreatment for Biofuels Production*,
SpringerBriefs in Green Chemistry for Sustainability,
DOI:10.1007/978-94- 007-6052-3_4,© The Author(s)2013

4.1 Introduction

Processing of lignocellulosic biomass requires breaking down the coherent structure in order to access all the needed components. The cellulose component needs to be hydrolyzed to fermentable glucose for the production of ethanol. Most of the cellulose is not accessible to enzymes because of its association with other lignocellulosic components. Non-treated lignocellulosic biomass usually produces glucose yields below 20 % (Zheng et al. 2009). Therefore, a suitable pretreatment has to be applied prior to hydrolysis in order to cleave the bonds among and within cellulose, hemicellulose, and lignin, as well as to increase its surface area and reduce the degree of polymerization (DP) and crystallinity.

The research performed to date shows that lignocellulosic biomass is potentially the most promising alternate source of energy and value-added products. It can be pretreated efficiently, cost-effectively, and in an environmentally friendly manner if the methods are correctly optimized. Ongoing research is focusing on optimizing and improving these technologies in order to reduce energy demands, the use of chemicals and the formation of by-products, and more importantly, to find applications for the coproducts produced during the lignocellulosic ethanol production process to create a complete and economically feasible biorefinery (Wyman 1996). A desired pretreatment method should be as simple as possible, applicable to a wide variety of feedstock, and should ensure purity of all products obtained (Chandra et al. 2007). Other important factors include catalyst use, catalyst recovery, and waste treatment (Zheng et al. 2009). Although currently there are no perfect pretreatment methods for lignocellulosic biomass, during the past decade advancements have been made in optimizing already discovered techniques and modifying old industrial processes for new applications (ethanol fermentation and lignin utilization). Based on many studies completed, one of the methods that show especially high potential in future research is hydrothermal pretreatment. Hydrothermal pretreatment uses no chemicals (or small amounts of mineral acids/alkali as catalysts), generates highly digestible cellulose, requires relatively short treatment time and moderate energy use, and has low equipment capital costs. All of these features suggest that hydrothermal pretreatment has low environmental impact and is highly effective in the improvement of lignocellulose's digestibility. It shows promise as a robust and effective biomass processing method to produce a wide range of biobased products. It can be designed as a green pretreatment method for on-farm or mobile applications.

4.2 HydrothermalPr etreatmentI nfluence on the Lignocellulosic Biomass

The principle of hydrothermal pretreatment is that high temperature causes auto-ionization of water, generates hydrogen ions, and thus reduces the pH significantly (Aita and Kim 2010). Water at high temperatures generates enough hydrogen ions

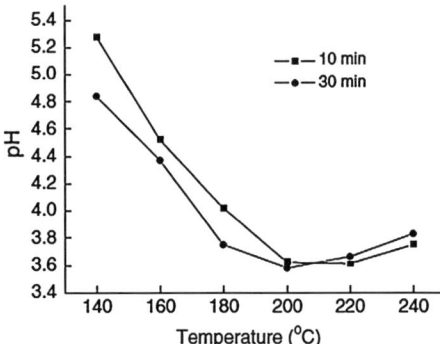

Fig. 4.1 pH relationships with temperature in the liquid fraction of the hydrothermally pretreated rice straw indicate pressure (Reprinted from Applied Biochemistry and Biotechnology, 160/2, Pretreatment of Rice Straw by a Hot-Compressed Water Process for Enzymatic Hydrolysis, Guoce Yu, Copyright 2008, with permission from Springer)

to drop the pH to acidic levels. At 200 °C the pH of water was found to be below 5 (Wyman et al. 2005). This process induces hemicellulose solubilization and hydrolysis of the acetyl groups. Acetic acid is a by-product of this reaction, which further catalyzes the hydrolysis. Hydronium ions generated from acetic acid are considered even more important than those of water origin (Garrote et al. 1999). Therefore, another name for the hydrothermal treatment is autohydrolysis (Aita and Kim 2010). The pH relationship with temperature in the liquid fraction of the hydrothermally pretreated rice straw can be found in Fig. 4.1 (Yu et al. 2010). This treatment has been proven to work especially well when applied to herbaceous materials and hardwoods, but was not efficient for softwoods due to the low contentofa cetylgroup sinthe s oftwoodhe micellulose(A lvirae ta l.2010).

Generally, hydrothermal pretreatment does not require size reduction prior to the process, especially when applied on a large scale, thus resulting in a large cost reduction (Taherzadeh and Karimi 2008). However, according to Hosseini and Shah (2009), 50 % increase in energy efficiency can be achieved when the size of wood chips is reduced. Hydrothermal pretreatment removes only small amounts of lignin (the acid soluble fraction), but it does change lignin structure by melting, coagulation, and its subsequent repolymerization on cellulose fibers. Therefore, it is not possible to extract lignin in its functional form from hydrothermally pretreated solids (Aita and Kim 2010). Changing the lignin structure results in cleavage of the linkages between lignin and carbohydrates. This opens the cellulose fibrils to the actions of enzymes. Syringil units in lignin were found to be most susceptible to degradation during hydrothermal pretreatment. Repolymerized lignin, referred to as "pseudolignin", contains residual xylans as well as hemicellulose degradation products, and is acid insoluble, thus resulting in false Klason lignin measurements of the solid fraction. Sugar degradation products tend to react with precipitating lignin, following the condensation route (Garrote et al. 1999; Young 1998). Repolymerization counteracts depolymerization and lignin removal, and although the ether bonds are being cleaved, delignification rates can be found close to zero (Li et al. 2007). Furthermore, as depicted in Li et al. (2007), the molecular weight of lignin measured in the pretreated material increases with the increasing severity factor, thus suggesting lignin's repolymerization.

Hydrothermal pretreatment greatly increases surface area of cellulose (by nonchemical swelling), which significantly enhances possible enzyme access (Chang et al. 1981; Sun and Cheng 2002) (Fig. 4.2). Due to its high severity of process conditions, this treatment is one of the methods creating high concentrations of sugar degradation products. These include furfural (from dehydration of pentoses) and 5-hydroxyl-methyl-furfural (from dehydration of hexoses). Both degradation products and acetic acid (along with small amounts of other organic acids, such as levulinic and formic acids) formed during the treatment inhibit yeast and other fermenting microorganisms when present in high concentrations.

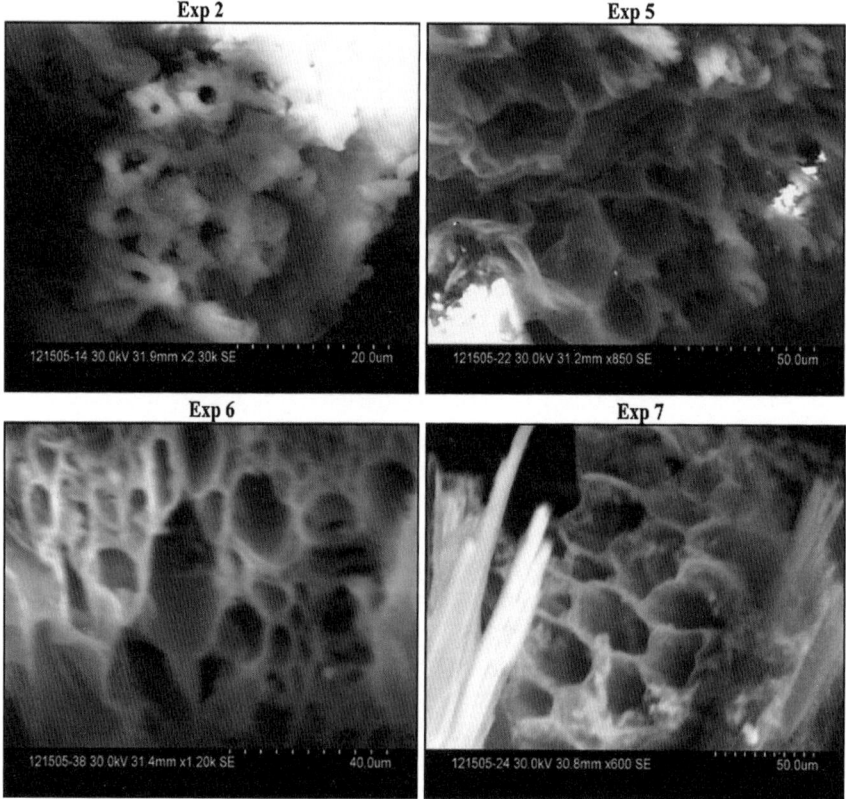

Fig. 4.2 Scanning electron microscopy (SEM) photographs of prairie cordgrass pretreated under various conditions: 210 °C/10 min (*Experiment* 2), 162 °C/15 min (*Experiment* 5), 218 °C/15 min (*Experiment* 6), and 190 °C/15 min (*Experiment* 7). (Reprinted with permission from Energy Fuels, Hydrothermal Pretreatment and Enzymatic Hydrolysis of Praire Cord Grass, Iwona Cybulska, Hanwu Lei, and James Julson, Copyright 2010, American Chemical Society)

Other by-products can include lignin degradation products, which are generally organic acids of phenylpropane origin (e.g., coumarilic acid, ferulic acid) (Aita and Kim 2010; Garrote et al. 1999). Hexose degradation to by-products become more favorable at temperatures above 210–220 °C (Garrote et al. 1999). There are methods for detoxification that could be applied in order to remove the inhibitory compounds, but it should be avoided due to the additional costs (Alvira et al. 2010). Instead, a simple filtration following the cooking step, would remove most of the by-products. Also, controlling pH during the process by addition of a base was found to reduce degradation of pentoses and therefore the formation of by-products (Taherzadeh and Karimi 2008). Steam explosion can also be part of hydrothermal pretreatment treatment with high-pressure saturated steam. The steam explosion process causes hemicelluloses degradation and lignin transformation due to high temperature (Sun and Cheng 2002). Limitations of steam explosion were reported, such as destruction of a portion of the xylan fraction, incomplete disruption of the lignin-carbohydrate matrix, and generation of compounds that might be inhibitory to microorganisms used in downstream processes (Mackie et al. 1985). Steam pretreated biomass needs to be washed with water to remove the inhibitory materials along with water soluble hemicellulose (McMillan 1994).

4.3 KineticsofHydr othermalPr etreatment (Time and Temperature Effects)

Generally, there are two models used for describing the kinetics of hydrothermal pretreatment: pseudohomogeneous first-order model and severity factor model (Garrote et al. 1999). The first model uses similar simplifications as the model of acid hydrolysis (the same mechanism of reaction) (Rodrigues Rde et al. 2010; Yat et al. 2008), and assumes that the polysaccharides hydrolysis follows irreversible and pseudohomogeneous first-order kinetics, with apparent coefficients following the Arrhenius equation for temperature dependence, and that hemicellulose degradation is independent of other processes occurring in the treatment. This model also neglects time, particle size, and pH effects (Saeman 1945). However, this model does not include the subsequent steps of the hydrothermal pretreatment, which are: diffusion of the hydronium ions into biomass structure, protonation of ether bonds, cleavage of ether bonds, and diffusion of the reaction products into the liquid phase (Garrote et al. 1999).

Severity factor is defined by an empirical equation, which is related to the most important factors of the treatment—time and temperature (Eq. 4.1) (Aita and Kim 2010;H endriksa ndZe eman2009):

$$R_0 = t e^{(T-100)/14.75} \qquad (4.1)$$

T—temperature, °C
t—time, min
R_0—severity factor

Its logarithmic form has been used by other authors (Eq. 4.2) (Galbe and Zacchi2007; H endriksa ndZe eman2009):

$$Log(R_0) = \log\left(t \, \exp\left(\frac{T - 100}{14.75}\right)\right) \qquad (4.2)$$

The pH influence can be also included in the equation (Eq. 4.3) (Galbe and Zacchi2007):

$$CS \text{ (combined severity)} = \log(R_0) - pH \qquad (4.3)$$

The optimal value of the severity factor for maximum enzymatic hydrolysis yield should be contained between 3.0 and 4.5 (Aita and Kim 2010). The relationship between the severity factor and sugar removal to the liquid fraction (biomass solubilization) can be found in Fig. 4.3a, while Fig. 4.3b presents the severity factor influence on the solid fraction yield remaining after the pretreatment (Ruiz et al. 2011).

Fig.4.3 Severity factor versus solubilized sugar concentration (**a**) and solid fraction yield (**b**) for wheat straw. Different *dots* represent various particle sizes of wheat straw blends: 0.488 mm, 0.330 mm, and 0.435 mm (Reprinted from Journal of Chemical Technology & Biotechnology, Hector A. Ruiz,D eniseS. Ruzene, Daniel P. Silva, Mafalda A.C.Q uintas, Antonio A. Vicente,J ose A. Teixeira, 88-94, Copyright 2010, with permission from Springer Science + Business Media)

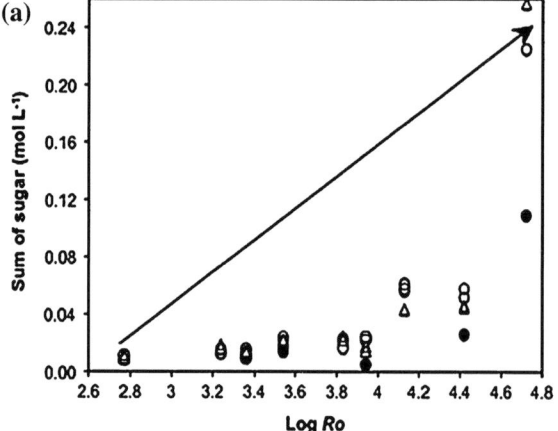

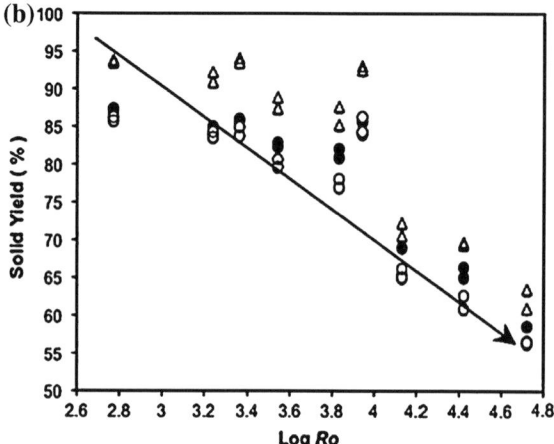

Severity factor equation can be used for each of the subsequent steps of the pretreatment process. Since hydrothermal treatment removes most of the hemicellulose, describing its kinetics can be based on the xylan removal from biomass using a following differential equation (Eq. 4.4) (which assumes first-order relation):

$$\frac{d\,(PX)}{dR_0} = -\,K\,(PX)$$

(4.4)

where:

PX—percent of xylan remaining in the solid, %

R_0—severity factor

K—kinetic constant independent from the severity of the process, min^{-1}

4.3.1 Influence of the Temperature Parameter

According to many studies performed to date, temperature has a major influence on the hydrothermal pretreatment efficiency and it determines the severity of the treatment more than the time factor (being in the exponent of the severity factor equation). The relationship between severity factor and temperature has been presentedinFig. 4.4(Y ue ta l.2010).

Temperatures below 180 °C were found to give low enzymatic hydrolysis glucose yields when applied to rice straw as a feedstock according to Yu et al. (2010) and Zhang et al. (2011). However, according to Cybulska et al. (2009), temperatures above 200 °C were found to produce high enzymatic hydrolysis glucose yields when hydrothermal pretreatment was applied to prairie cordgrass. These trends are presented in Figs. 4.5, 4.6, and 4.7 (Cybulska et al. 2009; Yu et al. 2010; Zhang et al. 2011). One of the reasons for an increased cellulose digestibility with increased pretreatment temperature is reduced DP of the cellulose fibers in samples treated at higher temperatures (Fig. 4.8). Generally, xylose yield rapidly decreases after a pretreatment at a temperature higher than 170 °C,

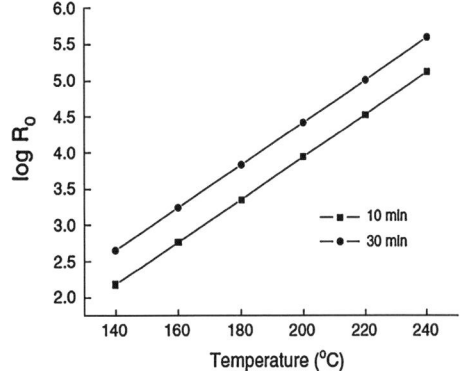

Fig.4.4 Severity factor versus temperature of the hydrothermal pretreatment (based on severity factor model in Eq. (4.2) (Reprinted from Applied Biochemistry and Biotechnology, 160/2, Pretreatment of Rice Straw by a Hot-Compressed Water Process for Enzymatic Hydrolysis,G uoce Yu, Copyright 2008, with permission from Springer)

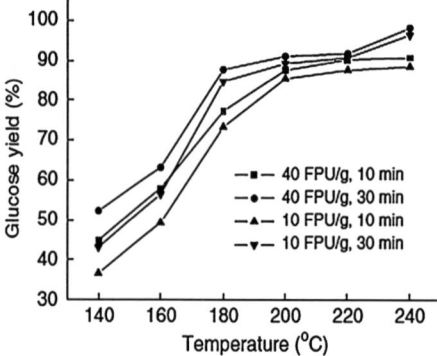

Fig. 4.5 Temperature influence on the enzymatic hydrolysis glucose yield of the solid fraction obtained after hydrothermal pretreatment of rice straw. (Reprinted from Applied Biochemistry and Biotechnology, 160/2, Pretreatment of Rice Straw by a Hot-Compressed Water Process for Enzymatic Hydrolysis, Guoce Yu, Copyright 2008, with permission from Springer)

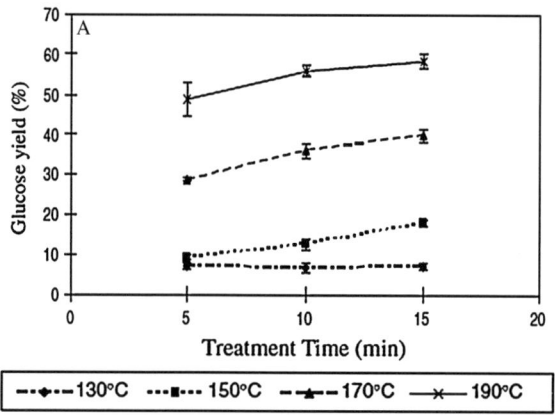

Fig.4.6 Temperature influence on the enzymatic hydrolysis glucose yield of the solid fraction obtained after hydrothermal pretreatment of cattails. (Reprinted from Journal of Industrial Microbiology and Biotechnology, 38/7, Hot water pretreatment of cattails for extractions of Cellulose, Bo Zhang, Copyright 2010, with permission from Springer)

as the degradation of pentoses becomes favorable. This relationship has been presented in Figs. 4.9, 4.10, and 4.11 (Cybulska et al. 2009; Yu et al. 2010; Zhang et al. 2011).

4.3.2 Influence of the Time Parameter

Based on the severity factor theory of hydrothermal pretreatment, time has a much lower influence on the pretreatment's effectiveness in improving cellulose digestibility than temperature. This fact has been proven by many experimental data, in

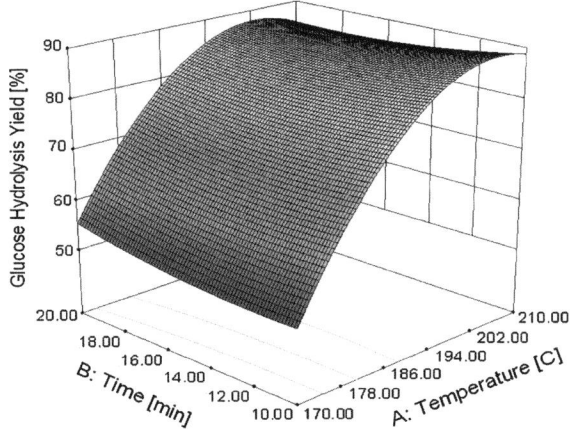

Fig.4.7 Temperature versus time versus glucose hydrolysis yield for hot water pretreatment of prairie cordgrass. (Reprinted with permission from Energy Fuels, Hydrothermal Pretreatment and Enzymatic Hydrolysis of Praire Cord Grass, IwonaC ybulska,H anwuL ei,a ndJ amesJ ulson, Copyright 2010, American Chemical Society)

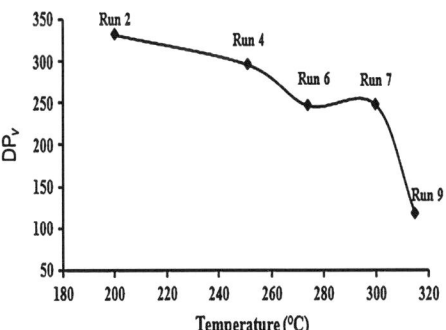

Fig. 4.8 Degree of polymerization of cellulose versus pretreatment temperature (Reprinted from Bioresource Technology, 101/4, Cellulose pretreatment in subcritical water: Effect of temperature on molecular structure and enzymatic reactivity, Sandeep Kumar, Rajesh Gupta, Y.Y. Lee, Ram B. Gupta, 1337-1347, Copyright 2010, with permission from Elsevier)

which time is mostly a factor of low significance in modeling of the sugar yields responses. The trend has been presented in Figs. 4.5, 4.6, 4.7, and 4.12 for the enzymatic hydrolysis glucose yield for different biomasses. Furthermore, the same trend was observed for xylose yields in both enzymatic hydrolysis and pretreatment liquid fraction (Figs. 4.9, 4.10, and 4.11). Similarly, experimental data show that inhibitory by-products formation does not depend strongly on the processing time,w hichis de pictedinFig. 4.13.

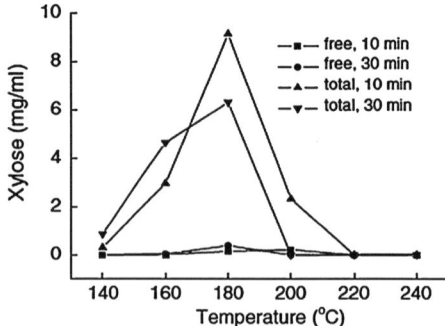

Fig. 4.9 Temperature influence on the xylose yield extracted to the liquid fraction (free means xylose in monomeric form; total includes both monomeric and oligosaccharide forms) of the hydrothermally pretreated rice straw. (Reprinted from Applied Biochemistry and Biotechnology, 160/2, Pretreatment of Rice Straw by a Hot-Compressed Water Process for Enzymatic Hydrolysis, Guoce Yu, Copyright 2008, with permission from Springer)

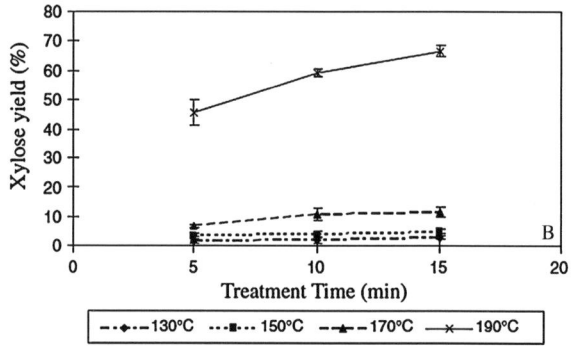

Fig. 4.10 Temperature influence on the enzymatic hydrolysis xylose yield of the solid fraction obtained after hydrothermal pretreatment of cattails. (Reprinted from Journal of Industrial Microbiology and Biotechnology, 38/7, Hot water pretreatment of cattails for extractions of Cellulose, Bo Zhang, Copyright 2010, with permission from Springer)

4.3.3 Influence of Other Factors (Biomass Loading and Pressure)

According to Taherzadeh and Karimi (2008) solids loading for the process ranges from 1 to 8 %. However, most of the industrial process conditions include solids loading of 10 % (liquor to solids ratio (LSR) is equal to 10) (Garrote et al. 1999). Pressure influence is directly related to the processing temperature used, and thus the relationship is similar (based on vapor pressure of water related to temperature). Enzymatic hydrolysis yield of the pretreated lignocellulosic materials tends to increase in the range between 0.5 and 2 MPa, then drastically decreases

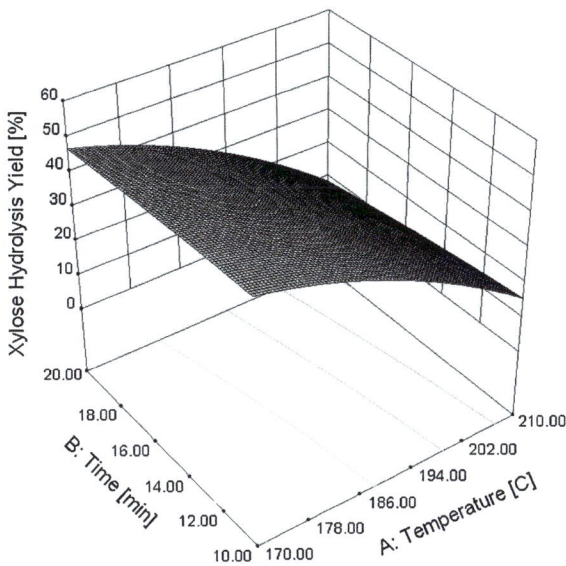

Fig. 4.11 Temperature and time influence on the enzymatic hydrolysis xylose yield from prairie cordgrass. (Reprinted with permission from Energy Fuels, Hydrothermal Pretreatment and Enzymatic Hydrolysis of Praire Cord Grass, IwonaC ybulska,H anwuL ei,a ndJ amesJ ulson, Copyright 2010, American Chemical Society)

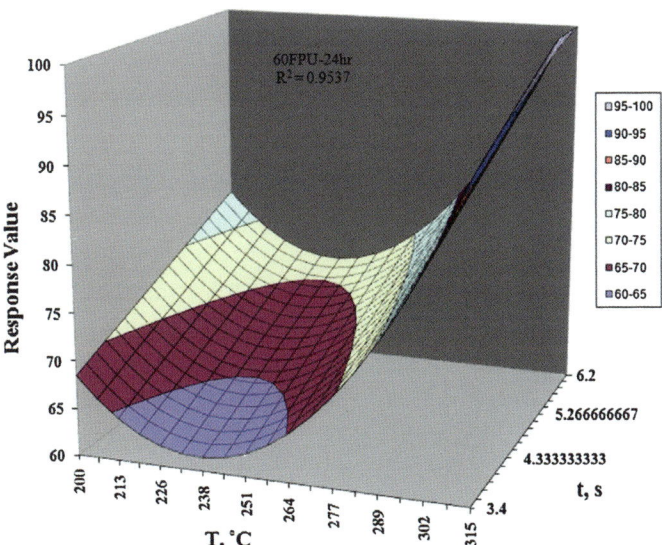

Fig. 4.12 Enzymatic hydrolysis glucose yield percentage (*response value*) after subcritical water pretreatment of cellulose (Reprinted from Bioresource Technology, 101/4, Cellulose pretreatment in subcritical water: Effect of temperature on molecular structure and enzymatic reactivity, Sandeep Kumar, Rajesh Gupta, Y.Y. Lee, Ram B. Gupta, 11 pages, Copyright 2010, with permission from Elsevier)

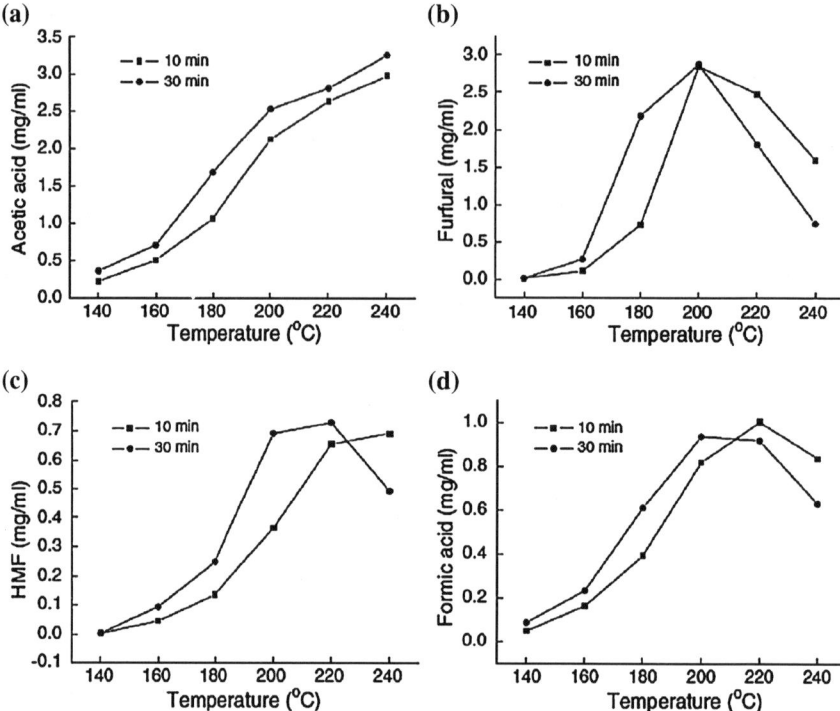

Fig. 4.13 Formation of inhibitory by-products in liquid fraction during hydrothermal pretreatment of rice straw. (Reprinted from Applied Biochemistry and Biotechnology, 160/2, Pretreatment of Rice Straw by a Hot-Compressed Water Process for Enzymatic Hydrolysis, Guoce Yu, Copyright 2008, with permission from Springer)

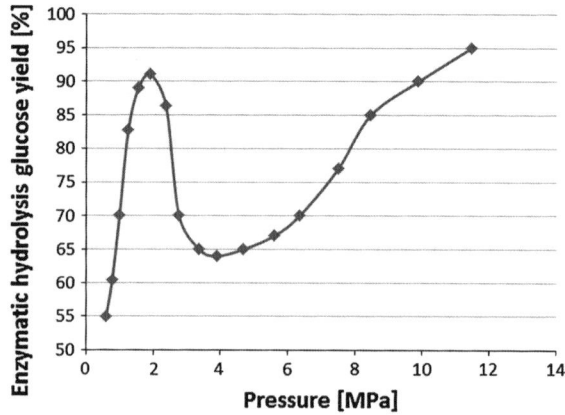

Fig. 4.14 Pressure influence on the enzymatic hydrolysis glucose yield of the lignocellulosic feedstock (combined data from pretreatment of Avicel cellulose and prairie cordgrass) (Cybulska eta l.2009; G eankoplis 2003; Kumar et al. 2010)

between 2 and 4 MPa, to eventually increase beyond 4 MPa (entering subcritical region for water). An exemplary graph combining two pressure regions from two researchs tudiesis pre sentedinFig. 4.14.

4.4 CharacteristicsofHydr othermalPr etreatment and Examples of Applications

4.4.1 Hot WaterPr etreatment

It has been established that hydrothermal pretreatment does not have to involve explosion, but the reactor is slowly heated and then cooled after the process (can be referred to as hot water treatment or liquid hot water cooking). The process is gaining interest as a pretreatment method for the ethanol industry since it does not require any chemicals and is simple in operation. It has been implemented on a pilot scale at DONG Energy facility in Skærbek, Denmark. Hot water pretreatment has been applied in a screw-conveying reactor. A particle pump prevents the biomass from splashing due to pressure release while exiting the reactor. Splashing is a common problem occurring at the location of biomass removal from the reactor (in either continuous or batch mode) (Petersen et al. 2009). The process schematic is presented in Fig. 4.15. Hot water pretreatment may be followed by a liquid and solid fractionation separation step, or the effluent can be fed into the next step in the form of a slurry (Wyman 1996). Hot water pretreatment can be applied to almost any type of lignocellulosic biomass. A study using olive tree residues revealed that the highest

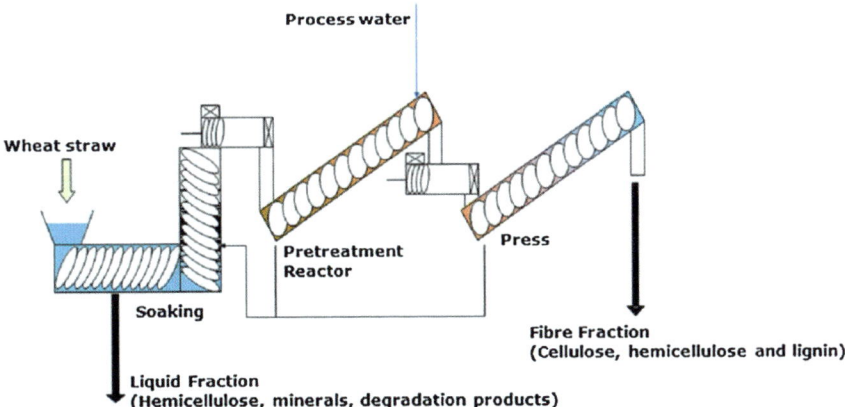

Fig. 4.15 Flow sheet a hydrothermal pretreatment process. (Reprinted from Biomass and Bioenergy, 33/5, Optimization of hydrothermal pretreatment of wheat straw for production of bioethanol at low water consumption without addition of chemicals, Mai Østergaard Petersen, Jan Larsen, Metter Hedegaard Thomsen, 834-840, Copyright 2009, with permission from Elsevier)

glucose yields obtained in enzymatic hydrolysis using the pretreatment temperatures between 200 and 210 °C are the most efficient, while the highest xylose recovery occurred at lower temperatures (~170 °C) (Cara et al. 2007). Glucose yield of 90 % occurred during enzymatic hydrolysis of hydrothermally pretreated (at 240 °C) yellow poplar sawdust. However, the inhibitory compounds formed during the treatment resulted in a low ethanol yield (50 %) (Weil et al. 1997).

According to Mok and Antal (1992), the efficiency of the treatment does not depend on the reaction conditions, but mainly on the feedstock type, especially when woody and herbaceous materials are being compared. However, the variability among the feedstock types was not found to be significant. It was also observed that compressed hot water percolation can remove 76–100 % of hemicellulose (depending on the material type) and up to 60 % of lignin at a severity parameter (log R_o) of 4.1 (at 230 °C for 2 min). Fermentation of the pretreated samples was not performed (Mok and Antal 1992). Good hemicellulose solubility (64 %) was also achieved in a study using hydrothermally treated wheat straw (with severity factor equal to 3.96, corresponding to 215 °C). Hemicellulose removal resulted in enriching the solids in cellulose up to 61 %. The amount of inhibitors generated was acceptable (Carvalheiro et al. 2009).

The most beneficial configuration for the hot water pretreatment reactor would be a continuous flow-through. This would ensure constant removal of hemicellulose from the material (Alvira et al. 2010). Other configurations include cocurrent and countercurrent flow reactors (Mosier et al. 2005b).

Hydrothermal pretreatment possesses some desirable features for bioethanol production. It is usually short (minutes), does not require any chemicals, and produces good hydrolysis and fermentation results. It cleaves the bonds between lignin and carbohydrates, and alters lignin structure to the point where it does not interfere with the enzymes. Moderate energy demand and water usage are also favorable characteristics of this green pretreatment method.

4.4.2 SteamE xplosion

Steam explosion is a type of hydrothermal pretreatment and one of the most common and efficient methods of lignocellulosic biomass pretreatment. In this process, steam is injected into a reactor along with biomass feedstock, resulting in swelling of the lignocellulosic structure. Steam temperature, which also generates high pressure is usually in the range of 180–260 °C (corresponding pressure, 0.69–4.83 MPa). After a specified reaction time, usually between 5 and 30 min, the steam is released suddenly through a release valve, causing the biomass structure to explode. The explosion rapidly disrupts the linkages between cellulose, lignin, and hemicellulose.

During steam explosion, some of the hexoses and pentoses from hemicellulose fraction are degraded to aldehydes and organic acids due to high temperature and pressure, which are inhibitory to fermenting microorganisms (Kosarie et al. 2001). The treatment does not involve any chemicals and has a moderate energy demand (Chandra et al. 2007). Steam explosion efficiency is affected by particle size of the feedstock (Zheng et al. 2009). For example, a steam explosion at 210 °C and 4 min residence time applied to poplar resulted in a 60 % glucose enzymatic hydrolysis yield, 60 % ethanol yield obtained during the simultaneous saccharification and fermentation (SSF), and 41 % xylose recovery in the liquid fraction (Negro et al. 2003). This pretreatment method is one of the few that have been demonstrated on both a pilot scale and a commercial scale (Wyman 1996; Zheng et al. 2009).

One of the industrial demonstration-scale steam explosion facilities is operated by Iogen Corporation in Canada. Its production capacity is 340 L of ethanol per ton of fiber (Zheng et al. 2009). Another example of the industrial-scale steam explosion pretreatment is the Masonite batch process used for production of fiber-board and other products in the early twentieth century (Mosier et al. 2005b). This pretreatment was applied to wood chips and used steam at pressures up to 90 atm and residence time between 1 and 10 min. A continuous mode of the Masonite process, called Stake II, uses an extruder as the reactor. Another example of a continuous industrial-scale process is the Rapid Steam Hydrolysis (RASH). In this process, the liquid fraction is continuously drained from the reactor, which reduces generation of inhibitory by-products (Garrote et al. 1999; Young 1998). A summary of hydrothermal pretreatment results for various biomass feedstocks is presented in Table 4.1.

4.4.3   CatalyzedHy drothermalPr etreatment

4.4.3.1 Sulfurdioxide C atalyzedHydr othermal Treatment

When compared to sulfuric acid impregnation, sulfur dioxide application creates less corrosion concerns, reduces gypsum formation, and produces higher xylose

Table4.1 Exemplary hydrothermal pretreatment results for various biomass feedstocks

Biomass	Typeof pretreatment	Optimal temperature (°C)/ time (min)/ pressure (MPa)	Enzymatic hydrolysis glucose yield (%)	Reference
Yellowpopla r sawdust	Hotw aterc ooking	230/<1/2.8	90	Weile ta l.(1997)
Prairiec ordgrass	Hotw aterc ooking	210/10/1.9	95	Cybulskae ta l. (2009)
Olivetre e	Hotw aterc ooking	210/60/1.9	76	Carae ta l.(2007)
Wheats traw	Hotw aterc ooking	195/6-12/1.4	93-94	Petersene ta l. (2009)
Cornfi ber	Hotw aterc ooking	260/0.17/4.7	100	Weil eta l.(1998)
Corns tover	Hotw aterc ooking	190/15/1.3	90	Mosiere ta l. (2005a)
Poplar	Steame xplosion	210/4/1.9	60	Negroe ta l. (2003)
Wheats traw	Steame xplosion	200/10	75	Ballesteros eta l.(2006)

yields (Aita and Kim 2010; Wyman 1996). As an example, SO_2 can be applied at the concentration of 1–4 % w/w biomass at typical steam explosion temperature and time (Taherzadeh and Karimi 2008). Catalyst application in hydrothermal treatment generally reduces the temperature and time needed for the autohydrolysis of lignocellulose, and also partially hydrolyzes cellulose (Aita and Kim 2010; Chandra et al. 2007). It is especially effective when applied to softwoods (Alvira et al. 2010).

A study with a two-step process of SO_2- catalyzed steam explosion applied to softwood showed high cellulose digestibility and ethanol yield (higher than a one-step process), achieving glucose yield of 80 % and ethanol yield of 69 % (Söderström et al. 2002). This process has also been practiced as a pretreatment method applied to agricultural residues with high efficiencies. Complete glucose to ethanol conversion and xylose to ethanol conversion was achieved by sugarcane bagasse pretreated by SO_2- impregnated steam explosion, utilizing recombinant yeast strains (Rudolf et al. 2008). High enzymatic hydrolysis glucose yields (~90 %) were also obtained from sweet sorghum bagasse treated by SO_2- impregnated steam explosion at ~200 °C for a short time (5–10 min) (Sipos et al. 2009). However, sulfur dioxide presents a health and safety hazard, and therefore the necessary precautions must be taken. There have been pilot-scale trials reported utilizing sulfur dioxide catalyzed steam explosion process (Wyman 1996; Zheng et al. 2009).

4.4.3.2 Carbondioxide C atalyzedHydr othermal Treatment

In this treatment, carbon dioxide under high pressure penetrates into the lignocellulosic structure, where it converts into carbonic acid, catalyzing the hydrolysis of hemicellulose. The studies using carbon dioxide include trials with supercritical carbon dioxide (Aita and Kim 2010). Supercritical carbon dioxide is in the form of a gas that has been compressed above the critical temperature and critical pressure, and therefore has some properties of a liquid. It is called a supercritical fluid. It possesses mass transfer properties of a gas and solvating power of a liquid. Carbon dioxide has been found to be a good lignin solvent, because it is nontoxic, nonflammable, and easy to recover (Alvira et al. 2010; Taherzadeh and Karimi 2008). There are several studies reporting application of carbon dioxide as a catalyst in the steam explosion process, most of them less efficient than Ammonia Fiber Explosion (AFEX) or the sulfur dioxide catalyzed process (Wyman 1996). However, according to Aita and Kim (2010), carbon dioxide use was found to be more economically feasible than AFEX when applied to agricultural and industrial residues.

Using supercritical carbon dioxide, which is a green solvent without the need for waste treatment, simultaneously with enzymes to hydrolyze lignocellulosic feedstock has been reported to result in high glucose yield (up to 100 %). Applying supercritical fluid during the hydrolysis can significantly increase the kinetic constants of the process, enabling the retention time to be shortened (Taherzadeh and Karimi 2008). Usage of different catalysts in the hydrothermal pretreatment process applied to various biomass feedstocks are presented in Table4.2.

Table4.2   Examples of catalyst uses for various types of biomass feedstocks

Biomass	Catalystt ype	Enzymatich ydrolysis glucose yield (%)	Reference
Softwood	Sulfurdi oxide	80	Söderströme ta l.(2002)
Sweets orghumb agasse	Sulfurd ioxide	90	Sipose ta l.(2009)
Switchgrass	Carbondi oxide	81	Luterbachere ta l.(2010)
Corns tover	Carbondi oxide	85	Luterbachere ta l.(2010)

4.4.4   HydrothermalPr etreatmentI ntegratedw ithO ther Methods

Due to drawbacks of individual pretreatment methods examined to date, many researchers have tried combining two or more pretreatment methods in series in order to enhance the overall performance. This includes combinations within the categories as well as between them. Generally, it has been found that integration of the treatment methods is more effective and selective than application of a single method; however, adding an extra step is usually associated with additional costs (Wyman1996).

An example of an integrated process is the combination of steam explosion and alkaline/hydrogen peroxide post-treatment. It was found that while steam explosion removes most of the hemicellulose component, the oxidative treatment that follows remove up to 80 % of the remaining lignin (Taherzadeh and Karimi 2008). Another example of an integrated biomass processing is an organosolv treatment followed by a hydrothermal post-treatment (Cybulska et al. 2012). The first step removes relatively pure lignin, which can be used as a value-added coproduct, while the hydrothermal post-treatment produces highly fermentable cellulose. It has been found that as much as 51 % of lignin can be recovered from prairie cordgrass and up to 67 % from corn stover, producing 79 and 90 % of enzymatic hydrolysis glucose yields for prairie cordgrass and corn stover, respectively (Cybulskae ta l. 2012).

4.5   Conclusion

Hydrothermal pretreatment is one of the most effective and environmentally friendly pretreatment methods available. It has been optimized for a wide variety of feedstocks and has been tested extensively throughout the years. The disadvantage of using high temperatures (and thus the need for pressure reactors and energy demand) is balanced with the advantages of no requirement of chemical additives and a short processing time. Excluding chemicals from the process reduces its environmental impact and eliminates the costs in the pretreatment and product recovery, making the hydrothermal pretreatment a rather "green" process.

Although high temperatures used during the hydrothermal processing favor formation of small amounts of organic acids and furans, these by-products can be converted into value-added products and used as an additional source of profit, or (especially in the case of acetic acid) recycled and utilized as an external catalyst of the lignocellulosic bonds cleavage.

As explained in this chapter, due to years of research, there are multiple versions of the hydrothermal pretreatment, thus there is a vast range of possibilities suitable for various settings. Depending on the biomass feedstock used, available resources, or application of the end product, the proper configuration of this process can be employed. It is one of the pretreatment methods that can be deployed on a form or on a mobile platform without the need for on-site waste treatment.

References

Aita G, Kim M (2010) Pretreatment technologies for the conversion of lignocellulosic materials to bioethanol. In: Sustainability of the sugar and sugar ethanol industries, vol 1058. American Chemical Society, Washington, D.C, USA, pp 117–145

Alvira P, Tomas-Pejo E, Ballesteros M, Negro MJ (2010) Pretreatment technologies for an efficient bioethanol production process based on enzymatic hydrolysis: a review. Bioresour Technol10 1(13):4851–4861

Ballesteros I, Negro MJ, Oliva JM, Cabañas A, Manzanares P, Ballesteros M (2006) Ethanol production from steam-explosion pretreated wheat straw. In: McMillan JD, Adney WS, Mielenz JR, Klasson KT (eds) Twenty-seventh symposium on biotechnology for fuels and chemicals, HumanaPre ss,pp 496–508

Cara C, Romero I, Oliva JM, Sáez F, Castro E (2007) Liquid hot water pretreatment of olive tree pruning residues. In: Mielenz JR, Klasson KT, Adney WS, McMillan JD (eds) Applied biochemistry and biotechnology, Humana Press, pp 379–394

Carvalheiro F, Silva-Fernandes T, Duarte LC, Girio FM (2009) Wheat straw autohydrolysis: process optimization and products characterization. Appl Biochem Biotechnol 153(1–3):84–93

Chandra R, Bura R, Mabee W, Berlin A, Pan X, Saddler J (2007) Substrate pretreatment: the key to effective enzymatic hydrolysis of lignocellulosics? In: Olsson L (ed) Biofuels, vol 108. Springer Berlin/Heidelberg, pp 67–93

Chang M, Chou T, Tsao G (1981) Structure, pretreatment and hydrolysis of cellulose. Bioenergy, 20:15–42 (Springer Berlin/Heidelberg)

Cybulska I, Brudecki G, Lei H, Julson J (2012) Optimization of combined clean fractionation and hydrothermal treatment of prairie cord grass. Energy Fuels 26(4):2303–2309

Cybulska I, Lei H, Julson J (2009) Hydrothermal pretreatment and enzymatic hydrolysis of prairie cord grass. Energy Fuels 24(1):718–727

Galbe M, Zacchi G (2007) Pretreatment of lignocellulosic materials for efficient bioethanol production. In: Olsson L (ed) Biofuels, vol 108. Springer, Berlin, pp 41–65

Garrote G, Dominguez H, Parajo JC (1999) Hydrothermal processing of lignocellulosic materials. HolzR oh- Werks57(3):191–202

Geankoplis C (2003) Transport processes and separation process principles (includes unit operations), 4th edn. Prentice Hall Press, New Jersey

Hendriks ATWM, Zeeman G (2009) Pretreatments to enhance the digestibility of lignocellulosic biomass.B ioresour Technol100(1):10–18

Hosseini SA, Shah N (2009) Multiscale modelling of hydrothermal biomass pretreatment for chip size optimization. Bioresour Technol100(9):2621–2628

Kosarie N, Sukan-Vardar F, Pieper HJ, Senn T (2001) The biotechnology of ethanol. Classical andfuture a pplications. Wiley-VCH VerlagG mbH, Wien

Kumar S, Gupta R, Lee YY, Gupta RB (2010) Cellulose pretreatment in subcritical water: effect of temperature on molecular structure and enzymatic reactivity. Bioresour Technol 101(4):1337–1347

Li J, Henriksson G, Gellerstedt G (2007) Lignin depolymerization/repolymerization and its critical role for delignification of aspen wood by steam explosion. Bioresour Technol 98(16):3061–3068

Luterbacher JS, Tester JW, Walker LP (2010) High-solids biphasic CO2–H2O pretreatment of lignocellulosic biomass. Biotechnol Bioeng107(3):451–460

Mackie KL, Brownell HH, West KL, Saddler JN (1985) Effect of sulphur dioxide and sulphuric acid on steam explosion of aspenwood. J Wood Chem Technol 5:405–425

McMillan J D (1994) Pretreatment of lignocellulosic biomass. In: Himmel M E, Baker JO, Overend RP (eds) Enzymatic conversion of biomass for fuels production, American Chemical Society, Washington, pp 292–324

Mok WSL, Antal MJ (1992) Uncatalyzed solvolysis of whole biomass hemicellulose by hot compressedliquidw ater.I ndE ngC hemR es31(4):1157–1161

Mosier N, Hendrickson R, Ho N, Sedlak M, Ladisch MR (2005a) Optimization of pH controlled liquid hot water pretreatment of corn stover. Bioresour Technol 96(18):1986–1993

Mosier N, Wyman C, Dale B, Elander R, Lee YY, Holtzapple M, Ladisch M (2005b) Features of promising technologies for pretreatment of lignocellulosic biomass. Bioresour Technol 96(6):673–686

Negro MJ, Manzanares P, Ballesteros I, Oliva JM, Cabanas A, Ballesteros M (2003) Hydrothermal pretreatment conditions to enhance ethanol production from poplar biomass. ApplB iochemB iotechnol105–108(1):87–100

Petersen MØ, Larsen J, Thomsen MH (2009) Optimization of hydrothermal pretreatment of wheat straw for production of bioethanol at low water consumption without addition of chemicals. Biomass Bioenergy33(5):834–840

Rodrigues Rde C, Rocha GJ, Rodrigues D Jr, Filho HJ, Felipe MG, Pessoa A Jr (2010) Scale-up of diluted sulfuric acid hydrolysis for producing sugarcane bagasse hemicellulosic hydrolysate (SBHH).B ioresour Technol101(4):1247–1253

Rudolf A, Baudel H, Zacchi G, Hahn-Hägerdal B, Lidén G (2008) Simultaneous saccharification and fermentation of steam-pretreated bagasse using Saccharomyces cerevisiae TMB3400 and *Pichias tipitis* CBS6054. Biotechnol Bioeng99(4):783–790

Ruiz HA, Ruzene DS, Silva DP, Quintas MAC, Vicente AA, Teixeira JA (2011) Evaluation of a hydrothermal process for pretreatment of wheat straw—effect of particle size and process conditions. J Chem TechnolB iotechnol86(1):88–94

Saeman JF (1945) Kinetics of wood saccharification—hydrolysis of cellulose and decomposition of sugars in dilute acid at high temperature. IndE ngC hem37(1):43–52

Sipos B, Réczey J, Somorai Z, Kádár Z, Dienes D, Réczey K (2009) Sweet sorghum as feedstock for ethanol production: enzymatic hydrolysis of steam-pretreated bagasse. Appl Biochem Biotechnol153(1) :151–162

Söderström J, Pilcher L, Galbe M, Zacchi G (2002) Two-step steam pretreatment of softwood; impregnation for ethanol production. Appl Biochem Biotechnol 98–100(1):5–21

Sun Y, Cheng J (2002) Hydrolysis of lignocellulosic materials for ethanol production: a review. Bioresour Technol83(1):1–11

Taherzadeh MJ, Karimi K (2008) Pretreatment of lignocellulosic wastes to improve ethanol and biogas production: a review. Int J Mol Sci 9(9):1621–1651

Weil J, Sarikaya A, Rau S-L, Goetz J, Ladisch C, Brewer M, Hendrickson R, Ladisch M (1998) Pretreatment of corn fiber by pressure cooking in water. Appl Biochem Biotechnol 73(1):1–17

Weil J, Sarikaya A, Rau SL, Goetz J, Ladisch CM, Brewer M, Hendrickson R, Ladisch MR (1997) Pretreatment of yellow poplar sawdust by pressure cooking in water. Appl Biochem Biotechnol68(1–2) :21–40

Wyman CE (1996) Handbook on bioethanol: production and utilization. Taylor & Francis, Washington

Wyman CE, Dale BE, Elander RT, Holtzapple M, Ladisch MR, Lee YY (2005) Coordinated development of leading biomass pretreatment technologies. Bioresour Technol 96(18):1959–1966

Yat SC, Berger A, Shonnard DR (2008) Kinetic characterization for dilute sulfuric acid hydrolysis of timber varieties and switchgrass. Bioresour Technol 99(9):3855–3863

Young RA (1998) Environmentally friendly technologies for the pulp and paper industry. Wiley, New York

Yu G, Yano S, Inoue H, Inoue S, Endo T, Sawayama S (2010) Pretreatment of rice straw by a hot-compressed water process for enzymatic hydrolysis. Appl Biochem Biotechnol 160(2):539–551

Zhang B, Shahbazi A, Wang L, Diallo O, Whitmore A (2011) Hot-water pretreatment of cattails for extraction of cellulose. J Ind Microbiol Biotechnol 38(7):819–824

Zheng Y, Pan Z, Zhang R (2009) Overview of biomass pretreatment for cellulosic ethanol production.IntJ AgricB iolEng 2: 51–68

Chapter 5
Pretreatment of Lignocellulosic Biomass Using Supercritical Carbon Dioxide as a Green Solvent

Tingyue Gu

Abstract Only concerted efforts utilizing various forms of energy can relieve today's energy crunch that threatens world economy and stability. Renewable bio-energy is an integral part of the solution. Lignocellulosic biomass is attractive for bioenergy production because it is cheap and ubiquitous. Unlike corn, its use does not interfere with the human and farm animal food supply chain. Unfortunately, by nature's design, lignocellulosic biomass is recalcitrant. It is difficult and costly to release the fermentable sugars from lignocellulosic biomass for ethanol fermentation. Thus, pretreatment is necessary. In the production of lignocellulosic ethanol, the biomass pretreatment step is often the most difficult and expensive part of the entire process. Many pretreatment methods have been proposed in the literature. Some of them require harsh chemicals that are not suited for a mobile or on-farm biomass processing unit. Supercritical CO_2 (SC-CO_2) explosion pretreatment uses CO_2, which is a green solvent, to treat biomass prior to enzyme hydrolysis. In glucose fermentation for bioethanol production, each mole of ethanol is accompanied by one mole of CO_2 by-product. Some of the CO_2 can be used for biomass processing without a net increase in CO_2 emission into the atmosphere. SC-CO_2 can diffuse into the crystalline structure of cellulose. The subsequent explosion action weakens the biomass cell wall structure and increases accessible surface areas for cellulase enzymes. SC-CO_2 also introduces acidity in moist biomass that helps pretreatment. This chapter discussed various aspects of the SC-CO_2 explosion pretreatment of lignocellulosic biomass including corn stover, wood, and switchgrass. Operating conditions, glucose yields for different types of lignocellulosic biomass, and pretreatment mechanisms were investigated.

Keywords Biomass pretreatment • Mobile pretreatment • Supercritical CO_2 • Cellulose • Hemicellulose • Lignin

T. Gu (✉)
Department of Chemical and Biomolecular Engineering, Ohio University,
Athens, OH 45701, USA
e-mail: gu@ohio.edu

T. Gu (ed.), *Green Biomass Pretreatment for Biofuels Production*,
SpringerBriefs in Green Chemistry for Sustainability,
DOI: 10.1007/978-94-007-6052-3_5, © The Author(s) 2013

5.1 Introduction

The world's petroleum reserves are estimated to last only for another few decades (Demirbas 2007). Although coal is far more abundant, its carbon emission is 40 % more than that for oil and 80 % more than natural gas in power generation (Peterson and Hustrulid 1998). In the United States, environmental concerns are hampering shale gas exploration and production. All these common forms of fossil fuels will eventually run out in the foreseeable future. Renewable energy must supplement the increasingly large shortfall. Among various forms of renewable energy, bioenergy is a key component. Corn ethanol production has been growing rapidly. More land is being allocated to corn used for bioethanol production, which drives up food prices. Lignocellulosic ethanol is an attractive alternative to corn ethanol because it uses various types of biomass that are agricultural wastes or can be grown on poor soils that are not suitable for food crops. RFA (2011) reported that there are more than 20 demonstration and pilot-scale facilities as of 2011 utilizing a variety of different technologies to convert lignocellulosic biomass such as wood, grasses, corn cobs, sugar waste, as well as algae and even garbage into ethanol.

Lignocellulosic biomass such as wood, corn stover, and switchgrass contain lignin (15–25 % w/w), hemicellulose (23–32 %), and cellulose (38–50%) (Mamman et al. 2008). Cellulose is a glucan with a general molecular formula of $(C_6H_{10}O_5)_n$. Hydrolysis of cellulose produces glucose that is a fermentable sugar. Hemicellulose can be hydrolyzed to release xylose, arabinose, glucose, mannose, and galactose. The dominant hydrolysis product xylose is considered unfermentable because most microorganisms cannot utilize xylose. Lignin contains phenylpropane units that are cross-linked together with a variety of chemical bonds. It is often burned as a high-energy fuel. Its complex chemistry has a potential as a feedstock to produce complicated chemicals. By nature's design, lignocellulosic biomass is recalcitrant because cellulose, hemicellulose, and lignin are integrated together (Chap. 1). The release of glucose by cellulase enzyme is hindered by the protective sheath of lignin, the presence of hemicellulose around cellulose, and the crystallinity of cellulose (Laureano-Perez et al. 2005). Thus, pretreatment is necessary. Without effective pretreatment, cellulase enzyme hydrolysis of cellulosic biomass yields less than 20 % of fermentable sugar, whereas proper pretreatment improves the yields up to 90 % (Alizadeh et al. 2005). Lignocellulosic ethanol production typically involves four major steps: biomass pretreatment, biomass hydrolysis, fermentation of released sugars, and ethanol separation from fermentation broth. Yang and Wyman (2008) estimated that the pretreatment step accounts for as much as 20 % of the total cost for cellulosic ethanol production. Thus, a cost-effective pretreatment step is important toward the economic competitiveness and success of lignocellulosic ethanol.

There are quite a few pretreatment methods that have been developed for lignocellulosic biomass in the literature (Mosier et al. 2005; Wyman et al. 2005; Hendriks and Zeeman 2009). They include (1) mechanical pretreatment Artz et al. 1990; Karunanithy and Muthukumarappan 2011), (2) biological pretreatment (Keller et al. 2003; Wan and Li 2010); (3) dilute acid pretreatment (Lloyd and Wyman 2005; Wang et al. 2011), (4) alkaline pretreatment (Carrillo et al. 2005; Hu and Wen 2008); (5) ammonia fiber explosion pretreatment (Teymouri et al. 2004, 2005; Alizadeh et al. 2005), (6) steam explosion and hot water pretreatment (Kaar et al. 1998; Montane et al. 1998; Laser et al. 2002; Pérez et al. 2008), (7) SC-CO_2 explosion pretreatment (Zheng et al. 1998; Narayanaswamy vet al. 2011; Luterbacher et al. 2012), (8) ionic liquid (IL) pretreatment (Li et al. 2010; Simmons et al. 2010), etc. Each method has its pros and cons. A single method is not expected to be suitable for all types of lignocellulosic biomass. Some of the pretreatment methods use harsh chemicals that lead to waste treatment problems. Mechanical, biological, steam, and hot water pretreatment methods are obviously green treatment methods because they do not involve chemical additives. Pretreatment using green ionic liquid is also green. However, the costs of such liquids are still too high. SC-CO_2 pretreatment is also considered green because CO_2 is recognized as a green solvent. SC-CO_2 pretreatment conditions are achieved more easily than some other methods. Thus, this green treatment is potentially attractive for mobile and on-farm applications without waste treatment problems. This chapter discussed various aspects in SC-CO_2 explosion pretreatment of lignocellulosic biomass.

5.2 Rationalef orSe lectingC O_2 for Supercritical Fluid Pretreatment

Figure 5.1 shows the schematic of a pure component solid–liquid–vapor phase diagram. When its temperature and pressure both exceed its critical values, a supercritical state is achieved. This means the liquid and gas phases become indistinguishable. A supercritical fluid possesses properties of both liquid and gas. In the supercritical region, the fluid becomes a special solvent because it possesses gas-like viscosity and liquid-like density. Its gas-like viscosity and diffusivity allow it to penetrate into the small pores of lignocellulosic biomass more easily than other solvents. Table 5.1 shows that among several common chemicals proposed for supercritical pretreatment, CO_2 has the lowest critical temperature (31.0 °C) and its critical pressure (1071 psi) is lower than that of ammonia, methanol, and water, respectively. Ethanol and n-propanol have the critical pressure of 926 and 934 psi, respectively, in Table 5.1. The pressures are lower but their critical temperatures of 243 and 263.6 °C are very high. It should be pointed out that biomass is usually wetted before SC-CO_2 pretreatment for a better sugar yield. This means that the system liquid is a mixture of CO_2 with a small amount of water. Its supercritical behavior will deviate from that of pure CO_2 slightly. Typical SC-CO_2 pretreatment uses a temperature and pressure much higher than the critical temperature and pressure for pure CO_2(Zhe nge ta l.1998).

Fig.5.1 Pressure-
temperature phase diagram of
a typical fluid showing solid,
liquid, and gas phases and
the supercritical fluid region
above the critical temperature
and pressure

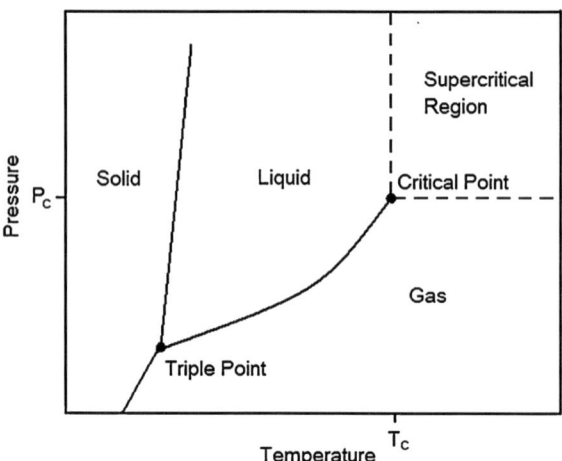

Table5.1 Critical temperature and pressure of some chemicals (Felder and Rousseau 2005)

Chemical	T_c (°C)	P_c(psi)
Carbondioxide (C O_2)	31.0	1071
Water(H_2O)	374.2	3208
Methanol(C H_3OH)	240.0	1154
Ethanol(C H_3CH_2OH)	243.1	926
n-Propanol(C H_3CH_2OH)	263.6	734
Ammonia(N H_3)	132.3	1636

Apart from a relatively easily achieved supercritical state, CO_2 has the following
additional advantages over other chemicals (Narayanaswamy 2010):

(1) itis n ontoxic(unlik ea mmonia),
(2) itis n onflammable(unlik ea lcohols),
(3) it does not cause degradation of sugars unlike steam explosion due to the
    higher temperature involved in the latter method (Zheng et al. 1995),
(4) ita ddsa ciditytomois tbioma sstha te nhancespre treatment,
(5) it can be transported in solid (dry ice), liquid, and gas forms,
(6) itis ine xpensive,a nd
(7) itis re adilya vailablefromthe e thanolfe rmentationproc ess.

The SC-CO_2 pretreatment method is environmentally friendly because CO_2
is considered a green solvent. Even if it is released to the atmosphere after use in
biomass pretreatment, it does not automatically mean an increase in the net CO_2
emission. In glucose fermentation for ethanol production, each mole of ethanol is
accompanied by one mole of CO_2 as a by-product. The produced CO_2 can be used
for biomass pretreatment before its release or sequestration. In fact, for cellulosic
ethanol product, the CO_2 originally comes from the atmosphere via photosynthesis.
Thus, its release is still carbon neutral. CO_2 is especially attractive for tactical

biomass processing because it is nontoxic, nonflammable, and inexpensive. Only a handful of papers in the open literature investigated this topic so far, probably because it is not favored for large-scale applications. Table 5.2 is a summary of SC-CO_2 pretreatment of various types of biomass reported in the literature that resulted from laboratory-scale investigations.

The first reported study was by Zheng et al. (1995) who explored the mechanism of SC-CO_2 explosion pretreatment on cellulose by using Avicel which is a commercially available pure form of cellulose. They subsequently tested the method on recycled paper and sugarcane bagasse with good glucose yields (Zheng et al. 1998). They also tested nitrogen, helium at 3000 psi and 35 °C and found that they also increased glucose yield, but to a lesser extent compared with CO_2. Kim and Hong (2001) investigated SC-CO_2 pretreatment on aspen and southern yellow pine wood and found that the method was effective on aspen, but not on pine. Gao et al. (2010) successfully used SC-CO_2 pretreatment on rice straw which is an abundant biomass in Asian countries. SC-CO_2 pretreatment was

Table 5.2 CO_2 explosion treatment of lignocellulosic biomass reported in the literature

Biomass	T(°C)	P(psi)	Time (min)	Maximum glucose yield	Reference
Avicel	35–80	1000–4000	60	0.74 g/ g Avicel	Zheng et al. (1995)
Avicel, recycled paper, bagasse	25–80	1100–4000	60	0.74 g/ g Avicel, 0.33 g/g recycled paper mix, 0.43 g/g bagasse	Zheng et al. (1998)
Aspen (hardwood), southern yellow pine (softwood)	112–165	3100 and 4000	10–60	84.7 % theoretical maximum for Aspen, 23 % for southern yellow pine	Kim and Hong (2001)
Rice straw	40–110	1450–4350	15–45	0.324 g/ g straw	Gao et al. (2010)
Guayule	100–200	2500–4000	30–60	77 % theoretical maximum	Srinivasan and Ju (2010)
Switchgrass, corn stover	160	2900	60	81 % for switchgrass and 85 % for corn stover (theoretical maximum)	Luterbacher et al. (2010)
Corn stover	80–150	3500	10–60	30 g/ 100 corn stover	Narayanaswamy et al. (2011)
Mixed hardwood, sunburst switchgrass	160–210	2900	60	83 % for hardwood and 80 % for switchgrass (theoretical maximum)	Luterbacher et al. (2012)

found effective for guayule which is a desert shrub used in the commercial production of hypoallergenic latex and resin constitutes (Srinivasan and Ju 2010). The commercial production process produces bagasse from guayule which can be a potential feedstock for lignocellulosic ethanol. Luterbacher et al. (2010) achieved high yields using SC-CO_2 pretreatment for switchgrass and corn stover. Narayanaswamy et al. (2011) successfully used liquid CO_2 in a siphoning type of CO_2 gas cylinder for a tubular reactor to achieve supercritical state without the need for a CO_2 pump to treat corn stover with high glucose yield.

Most recently, Luterbacher et al. (2012) used SC-CO_2 to treat mixed hardwood and Sunburst switchgrass with success. They recommended a two-temperature pretreatment strategy: 210 °C for 16 min followed by 150 °C for 60 min. Unlike other SC-CO_2 studies that focused only on glucose yield, they also investigated the effect of SC-CO_2 pretreatment on hemicellulose conversion. They found that SC-CO_2 not only enhanced glucose yield, but also xylose, arabinose, and mannose yields from hemicellulose after enzyme hydrolyses.

Apart from these SC-CO_2 explosion pretreatment cases, Muratov et al. (2005) tested cellulase enzymes from three different microorganisms (*Trichoderma viride, Trichoderma reesei, and Aspergillus niger*) by dissolving them in supercritical CO_2 (160 atm, 50 °C) mixed with a pH 5 acetate buffer for the hydrolysis of cotton fiber lasting 48 h. Improvement of glucose yield of only 20 % was observed compared with hydrolysis at 1 atm without CO_2. Their process did not take advantage of increased pore access due to the explosion of CO_2. They used supercritical CO_2 only as a special solvent.

5.3 SC-CO_2 Pretreatment Procedure

Narayanaswamy (2010) built a laboratory SC-CO_2 explosion device shown in Fig. 5.2. The main reactor tube had an outer diameter of one inch with a pressure rating of 4000 psi. Prior to SC-CO_2 pretreatment, the biomass was cut into small pieces (~1.2 mm) and soaked in water overnight. Five grams of wetted biomass was then treated with SC-CO_2. To measure the glucose yield versus untreated biomass, the pretreated biomass and untreated biomass were both hydrolyzed using cellulase enzyme and β-glycosidase enzyme to obtain glucose. Glucose yield was measured using a glucose kit (Narayanaswamy et al. 2011). For each gram of cellulose, its theoretical maximum glucose yield is 1.11 g, because each glucopyranose unit in cellulose polymer gains one water molecule after hydrolysis. The entire experimental flowchartis s howninFig. 5.3.

Figure 5.4 is a stirred tank reactor for SC-CO_2 pretreatment of lignocellulosic biomass at high solid loadings (Luterbacher et al. 2012). CO_2 was fed to the reactor after it was chilled. Narayanaswamy et al. (2011) used a siphoning type of CO_2 cylinder to feed liquid CO_2 into the SC-CO_2 explosion reactor without an expensive CO_2 pump. The CO_2 cylinder provided enough pressure at room temperature. The amount of CO_2 was controlled by measuring the CO_2 mass gained by the

Fig.5.2   Reactor used for the SC-CO$_2$ pretreatment (batch process) by Narayanaswamy (2010)

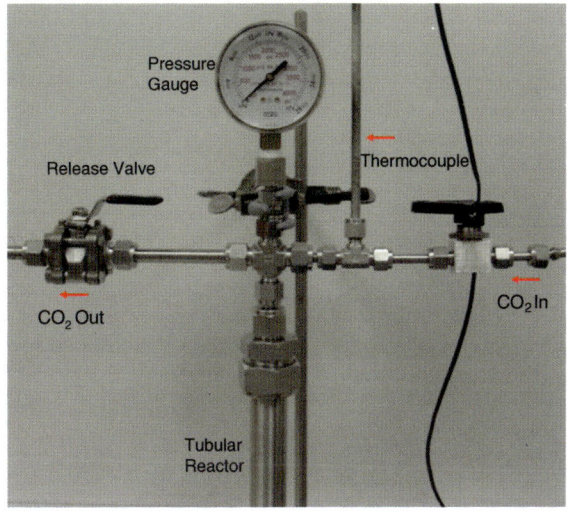

Fig.5.3   Flowchart of the experimental procedure used for pretreatment of the biomass(N arayanaswamy 2010)

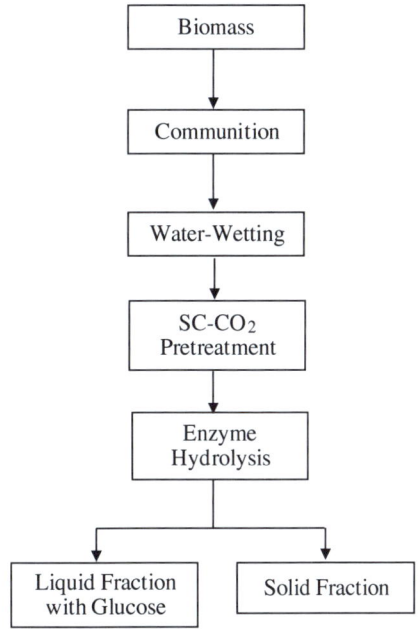

reactor. Supercritical CO$_2$ state was achieved by heating the reactor to a desired temperature. Increasing the reactor temperature increased CO$_2$ pressure as well. With a fixed amount of CO$_2$ for a fixed amount of moist biomass, a desired pressure was achieved at the set-point temperature. For higher temperatures, an oil or sand bath can be used in laboratory settings. In a real-world deployment, steam or electricalhe atingc anbe us ed.

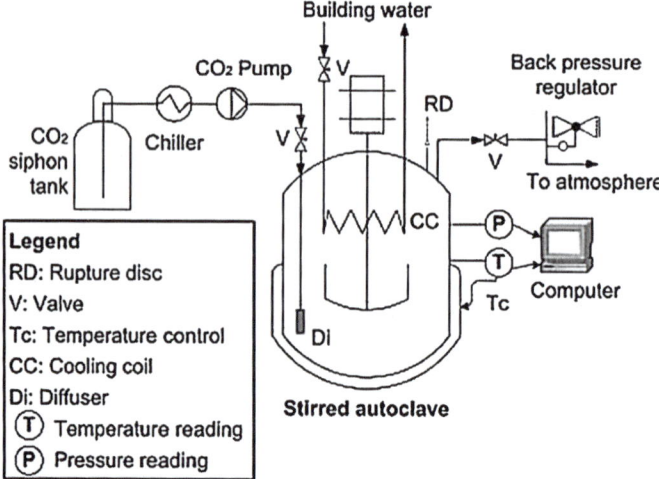

Fig. 5.4 Stirred tank reactor for SC-CO2 pretreatment of lignocellulosic biomass at high solid loadings (Reprinted from Biotechnology and Bioengineering, Two-temperature stage biphasic CO_2–H_2O pretreatment of lignocellulosic biomass at high solid loadings, Jeremy S. Luterbacher, Jefferson W. Tester, Larry P. Walker, Copyright 2012, with permission from John Wiley and Sons)

Fig. 5.5 Gas-liquid phase partitioning line gradually lowered toward the bottom of the glass window and eventually disappeared in the presence of wetted biomass when temperature increased from 12 °C to 40 °C in a pressurized chamber (Narayanaswamy2010)

Figure 5.5 shows a series of visual images on the transition from gas–liquid binary-phase CO_2 (with moist biomass in the liquid phase) to a single supercritical CO_2 phase (gas and liquid phase indistinguishable) observed through a glass window (Narayanaswamy 2010). At the initial temperature, a clear partitioning line between the gas and liquid phases is clearly visible in Fig. 5.5. Heating the CO_2 with submerged biomass raised the system temperature and pressure. The liquid phase volume shrank and became cloudy. The phase partition line eventually disappeared when the temperature was raised to 40 °C as shown in Fig. 5.5. Narayanaswamy (2010) suggested that the CO_2 dosage required for a certain amount of moist biomass to achieve a desired certain temperature and pressure could be predicted using a modified van der Waals equation of state or using experimental correlations from an SC-CO_2re actor.

5.4 Mechanisms

To probe possible mechanisms for the SC-CO_2 explosion effect on biomass, Zheng et al. (1995) used Avicel, a form of pure cellulose. They observed a 50 % reduction in cellulose crystallinity after SC-CO_2 pretreatment compared with untreated Avicel. They suggested that water molecules in general do not penetrate cellulose crystalline lattices. Their solid-state NMR data indicated a loosening of cellulose crystalline structure after SC-CO_2 pretreatment. Thus, they reasoned that SC-CO_2 was able to penetrate the cellulose crystalline structure.

Surprisingly, Narayanaswamy et al. (2011) found that SC-CO_2 made no change in crystallinity in corn stover. They argued that Avicel is a pure cellulose without other biopolymers in cell walls. In corn stover, however, cellulose microfibrils are embedded in hemicelluloses, lignin, and glycoproteins (Buchanan et al. 2000). This is somewhat consistent with data obtained by Kim and Lee (2005) who found that their ammonia pretreatment of corn stover showed undetectable change in the crystalline structure while the glucose yield was enhanced greatly.

It is reasonable to believe that SC-CO_2 explosion opened up small pores in the biomass and thus increased accessible areas for subsequent enzyme hydrolysis. Scanning Electron Micrograph (SEM) images shown in Figs. 5.6 and 5.7 support this argument. Another likely contributing factor could be that CO_2 dissolution in water increases acidity of the water that wets the biomass (Narayanaswamy et al. 2011; Zheng et al. 1995). Acid hydrolysis can dissociate cellulose-hemicelluloses-pectin network and this enhances cellulose hydrolysis. Figure 5.8 shows that the pH value of water-CO_2 system at 37 °C approaches 3 when pressure reaches 10 MPa and beyond (Meyssami et al. 1992). The pH effect theory is somewhat supported by tests with and without moisture performed by Narayanaswamy et al. (2011). They observed that without moisture, the SC-CO_2 pretreatment achieved only a very small improvement in glucose yield compared with a considerable increase in glucose yield when biomass was prewetted. Zheng et al. (1998) also pointed out the importance of water content in the biomass for SC-CO_2

Panel A   Panel B

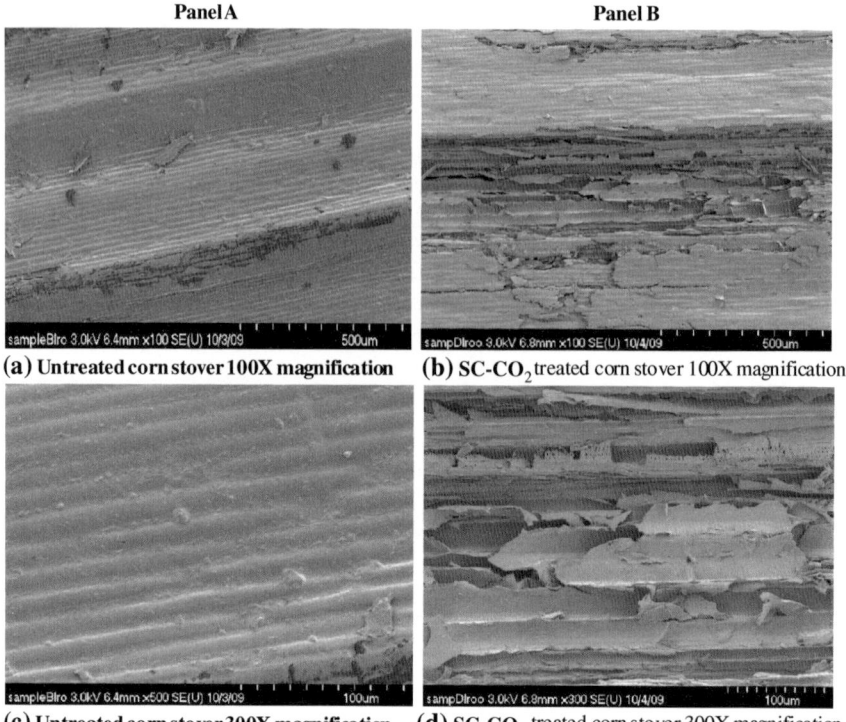

(a) Untreated corn stover 100X magnification   (b) SC-CO$_2$ treated corn stover 100X magnification

(c) Untreated corn stover 300X magnification   (d) SC-CO$_2$ treated corn stover 300X magnification

Fig. 5.6 SEM images of untreated and SC-CO$_2$ treated corn stover samples at 100X and 300X magnifications (Reprinted from Bioresource Technology, 102/13, Supercritical carbon dioxide pretreatment of corn stover and switchgrass for lignocellulosic ethanol production, Naveen Narayanaswamy, Ahmed Faik, Douglas J. Goetz, Tingyue Gu, 102(13):6995–7000, Copyright 2011, with permission from Elsevier)

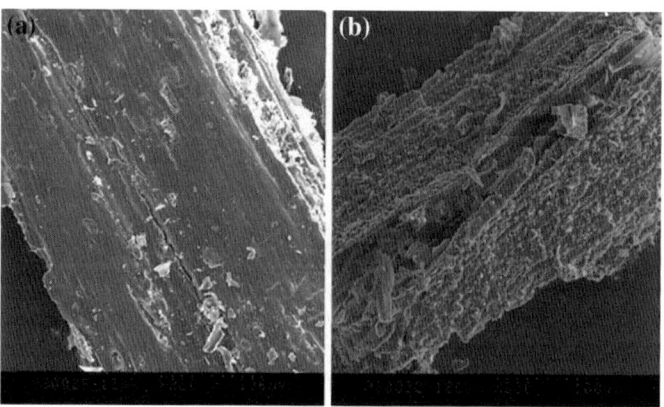

Fig. 5.7 SEM images showing the transverse section of rice straw before (**a**) and after (**b**) SC-CO$_2$ pretreatment. Reprinted from Biosystems Engineering, 106/4, Effect of SC-CO$_2$ pretreatment in increasing rice straw biomass conversion, Miao Gao, Feng Xu, Shurong Li, Xiaoci Ji, Sanfeng Chen, Dequan Zhang, 106:470–475, Copyright 2010, with permission from Elsevier

Fig.5.8 Measured and predicated pH of pure water-CO_2 simulation system at 37 °C temperature and pressure up to 34 MPa. (Reprinted from Biotechnology Progress, Prediction of pH in Model Systems Pressurized with Carbon Dioxide, Behrouz Meyssami,MuratO . Balaban,Arthur A. Teixeira, 8:149–154, Copyright 1992, with permission from John Wiley and Sons)

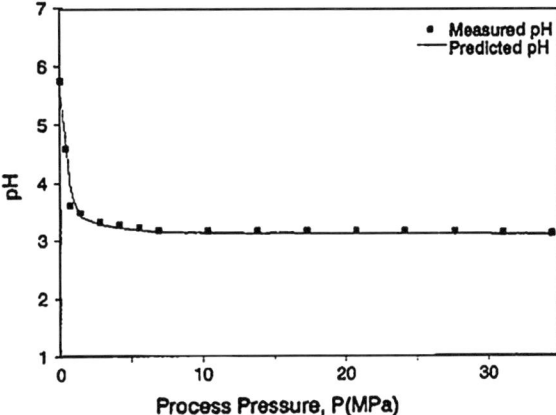

pretreatment because of carbonic acid formation. The pH introduced by CO_2 is less acidic than the pH in the range of 1.0–2.5 used by Wang et al. (2011) and Xu and Tschirner (2012) for dilute acid pretreatment. However, the acidity due to CO_2 comes at no cost and does not present a problem for wastewater because the acidity is gone when CO_2e scapesfromw ater.

5.5 PretreatmentP arameters

There are several key operating parameters for SC-CO_2 pretreatment because of the various mechanisms involved. Among them, temperature, pressure, treatment time, moisture content, CO_2/biomass ratio, are particularly worthy of investigations. So far, there is no detailed study on pressure release gradient.

5.5.1 EffectofT emperature

Zheng et al. (1998) studied the temperature effect on SC-CO_2 pretreatment of Avicel at 3000 psi. Their data in Fig. 5.9 show that at a subcritical temperature of 25 °C, there was only a small gain in glucose yield. A significant enhancement of glucose yield was observed when the temperature reached supercritical 35 °C. Further increasing the temperature to 80 °C saw only a very small additional gain in glucose yield. It should be noted that Avicel is a pure form of cellulose that is less recalcitrant than lignocellulosic biomass. Zheng et al. (1998) also noticed significant gains in glucose yield from sugarcane bagasse when the SC-CO_2 pretreatment temperature was raised from 35 to 80 °C at 1100 psi and 3000 psi, respectively. Their bagasse was easier to treat because it was dilute acid-hydrolyzed prior to the SC-CO_2 treatment.

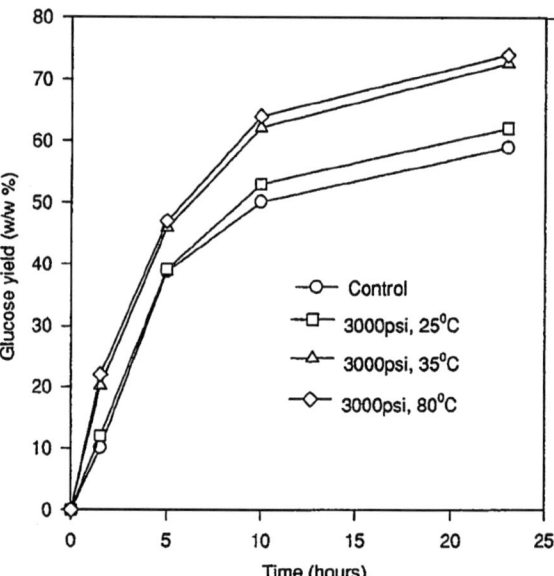

Fig.5.9 Temperature effecton Avicelus ing subcritical and super critical CO_2. (Reprinted from Biotechnology Progress, Pretreatment for Cellulose Hydrolysis by Carbon DioxideExplos ion, Yizhou Zheng, H.-M. Lin,G eorge T. Tsao,14: 890–896,C opyright 2008, with permission from John Wiley and Sons)

Kim and Hong (2001) found that at temperatures as low as 112 °C and 138 °C, the SC-CO_2 pretreatment at 3100 psi for 60 min had negligible effect on sugar yield. They recommended 165 °C that had a very large enhancement in sugar yield from aspen word. For rice straw treated by SC-CO_2 at 30 MPa for 30 min, Gao et al. (2010) found gradual improvements in glucose yield when the temperature was increased from 40 to 80 and then to 110 °C.

5.5.2 EffectofPr essure

Narayanaswamy et al. (2011) investigated temperature effect on SC-CO_2 pretreatment of corn stover. Their data at 3500 psi in Fig. 5.10 show that pretreatment at 80 °C for 60 min was inadequate, while 120 °C achieved a considerable enhancement in glucose yield. Increasing the pretreatment temperature to 150 °C further increased the glucose yield. Although the pretreatment temperature for lignocellulosic biomass favor a temperature much higher than the critical temperature, it is still much lower than the 250 °C used in steam treatment by Kim and Hong (2001).

A higher pressure helps the supercritical fluid to penetrate deeper and allows a more powerful explosion action to open up pores in the biomass. Figure 5.11 (Narayanaswamy et al. 2011) showed that 3500 psi achieved a twice as much enhancement as 2500 psi in SC-CO_2 pretreatment of corn stover at 150 °C for 60 min compared with untreated corn stover. Zheng et al. (1998) observed that there was an incremental increase in glucose yield from Avicel treated with SC-CO_2 at 35 °C when the 1100, 2000, 3000, 4000 psi were tested, respectively. Surprisingly, Gao et al. (2010) found that 20 MPa showed almost no improvement over 10 MPa

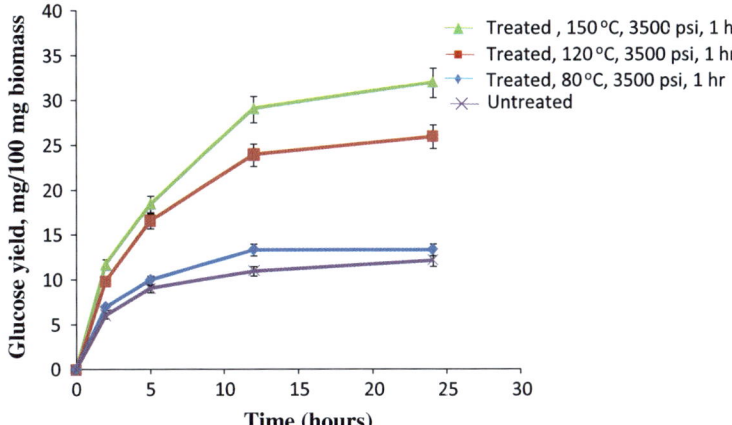

Fig. 5.10 Effect of temperature on SC-CO$_2$ pretreatment of corn stover (Reprinted from Bioresource Technology, 102/13, Supercritical carbon dioxide pretreatment of corn stover and switchgrass for lignocellulosic ethanol production, Naveen Narayanaswamy, Ahmed Faik, Douglas J. Goetz, Tingyue Gu, 102(13):6995–7000, Copyright 2011, with permission from Elsevier)

Fig.5 .11 Effect of pressure on SC-CO$_2$ pretreatment of corn stover. (Reprinted from Bioresource Technology, 102/13, Supercritical carbon dioxide pretreatment of corn stover and switchgrass for lignocellulosic ethanol production,N aveen Narayanaswamy, Ahmed Faik,D ouglasJ .G oetz, TingyueG u,102(13):6995–7000, Copyright 2011, with permission from Elsevier)

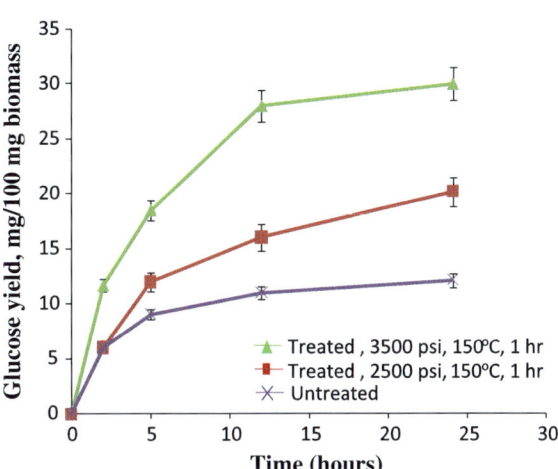

while 30 MPa showed a large improvement in glucose yield from rice straw treated with SC-CO$_2$ at 110 °C for 30 min. Figure 5.12 indicates that for aspen (hardwood) and southern yellow pine (softwood), 4000 psi actually had a lower sugary ieldc omparedw ith3 100p sia ccordingt oK ima ndH ong(2001).

5.5.3 EffectofPr etreatmentT ime

Zheng et al. (1995) used a fixed time of 1 h as SC-CO$_2$ pretreatment time for Avicel in the temperature range of 35–80 °C, and the pressure range of 1000–4000

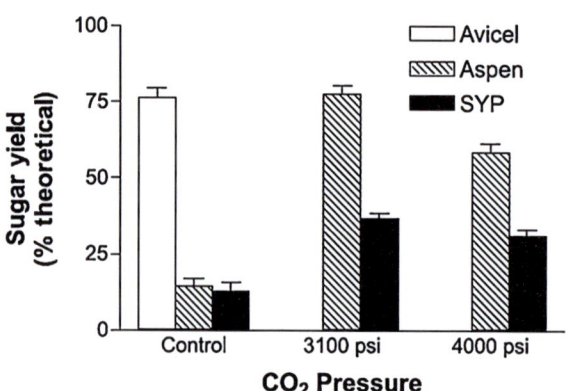

Fig.5.12 Effect of pressure on SC-CO_2 pretreatment of Avicel, Aspen,a nds outhern yellow pine. (Reprinted from Bioresource Technology, 77/2, Supercritical CO_2 pretreatment of lignocellulose enhances enzymatic cellulose hydrolysis,K young Heon Kim, Juan Hong, 139–144, Copyright 2001, with permission from Elsevier)

psi without doing a parametric study on time. It was found by Gao et al. (2010) that 30 min was sufficient for SC-CO_2 pretreatment of rich straw at 110 °C because 45 min did not improve glucose yield. Kim and Hong (2001) also concluded that 30 min was sufficient for SC-CO_2 pretreatment of aspen wood and southern yellow pine at 3100 psi and 165 °C because 60 min showed no improvement in sugar yield. Pretreatment time had a big impact on SC-CO_2 pretreatment of corn stover at 3500 psi and 150 °C as shown in Fig. 5.13 by Narayanaswamy et al. (2011). The figure indicates that 30 min was clearly not optimal because increasing the pretreatment time to 1 h more than doubled the yield enhancement over untreated corn stover. A pretreatment time of 16 min for hardwood (or 1 min for switchgrass) at 210 °C followed by 60 min at 150 °C was commended by Luterbachere ta l.(2012).

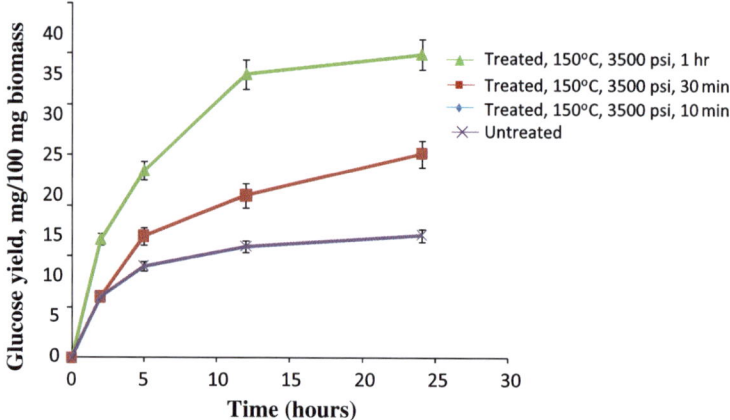

Fig. 5.13 Effect of time on SC-CO_2 pretreatment of corn stover. (Reprinted from Bioresource Technology, 102/13, Supercritical carbon dioxide pretreatment of corn stover and switchgrass for lignocellulosic ethanol production, Naveen Narayanaswamy, Ahmed Faik, Douglas J. Goetz, Tingyue Gu, 102(13):6995–7000, Copyright 2011, with permission from Elsevier)

5.5.4 EffectofM oisture

A small amount of moisture in the biomass appears to be essential for a successful pretreatment using SC-CO$_2$. SC-CO$_2$ pretreatment at 3500 psi, 120 °C for
60 min had only a small enhancement of glucose yield for dry corn stover while
75 % (w/w) moisture almost doubled the glucose yield (Narayanaswamy et al.
2011). This is consistent with the finding by Zheng et al. (1998) that there was
almost no improvement in glucose yield from SC-CO$_2$ pretreatment of dry Avicel.
They argued that water helped swell the pores in the biomass before the SC-CO$_2$
pretreatment. Narayanaswamy et al. (2011) also stressed the beneficial effect
of acidity provided by dissolved CO$_2$ in water as discussed early in Sect. 5.4.
Their data in Fig. 5.14 indicate that SC-CO$_2$ pretreatment had only a marginal
improvement on glucose yield for dry corn stover. A higher level of moisture
would likely not provide further enhancement for SC-CO$_2$ pretreatment (Zheng
et al. 1998). Kim and Hong (2001) tested moisture contents of 40, 57, 73 % (w/w),
respectively for SC-CO$_2$ pretreatment of aspen wood and southern yellow pine at
3100 psi and 165 °C for 30 min. They found that 57 % was sufficient and 73 % did
nots howa f urtheri mprovementi ns ugary ield(Fig.5.15).

5.5.5 CO$_2$ to Biomass Ratio

The amount of CO$_2$ for a fixed volume reactor containing biomass is largely deter-
mined by thermodynamics at the set-point operating pressure and temperature just
before explosion assuming that the biomass quantity change does not change the
fluid volume much. Biomass quantity can vary as long as the reactor volume is sufficiently large to hold the biomass and CO$_2$ and there is sufficient volume in the
reactor for fluid to reach an equilibrium at the set-point supercritical pressure and

Fig.5.14 Effect of moisture on SC-CO$_2$ pretreatment of corn stover. (Reprinted from Bioresource Technology, 102/13, Supercritical carbon dioxide pretreatment of corn stover and switchgrass for lignocellulosic ethanol production,N aveen Narayanaswamy, Ahmed Faik,D ouglasJ .G oetz, TingyueG u,102(13):6995–7000, Copyright 2011, with permission from Elsevier)

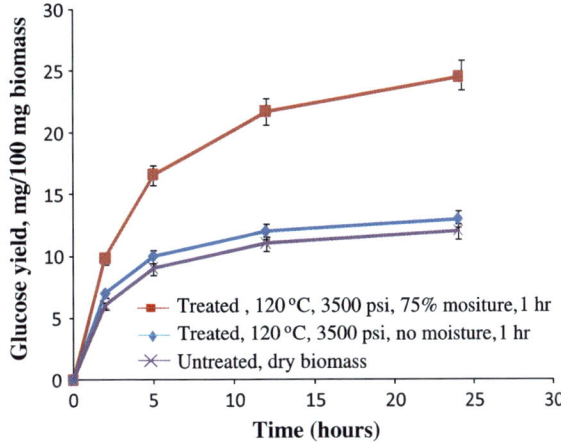

Fig.5.15 Effect of
moisture content on SC-CO$_2$
pretreatmentof Avicel,
Aspen, and southern yellow
pine. (Reprinted from
Bioresource Technology,
77/2, Supercritical CO$_2$
pretreatment of lignocellulose
enhances enzymatic cellulose
hydrolysis,K youngH eon
Kim, Juan Hong, 139–144,
Copyright 2001, with
permission from Elsevier)

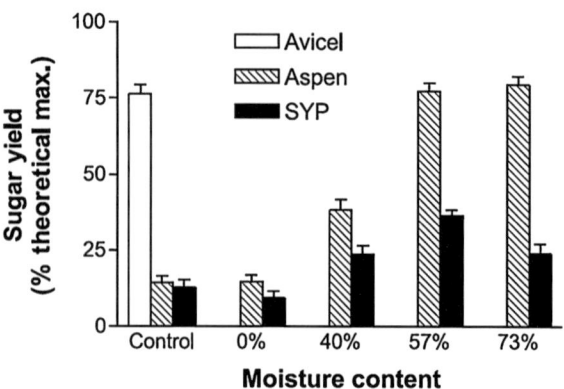

temperature. Narayanaswamy (2010) used a CO$_2$ to dry biomass loading ratio of roughly 10:1 (w/w) for SC-CO$_2$ pretreatment. This ratio is the same as those for aqueous ammonia, acid hydrolysis pretreatment methods (Narayanaswamy 2010), as well as the hydrothermal pretreatment method (Garrote et al. 1999). This ratio compares favorably with the 20:1 and 30:1 ratios commonly used for ionic liquid pretreatment (Table 6.4 in Chap. 6). It is not a minimized ratio. Only a small amount of CO$_2$ is needed to "soak" the entire biomass at the supercritical state. This ratio can be minimized to use less CO$_2$ if CO$_2$ is released into the air after SC-CO$_2$ pretreatment. It is also possible to design a more sophisticated process that recycles CO$_2$, but this would increase operating energy cost in addition to capital investment.

5.6 MobileandOn-far mPr etreatmento fB iomassU sing Green Chemistry

Biomass collection and transportation to a centralized facility are expensive. Biomass types such as corn stover and switchgrass are typically bulky. The end products of lignocellulosic ethanol production are ashes and other wastes that need to be disposed of. An alternative approach is to use small on-farm or even a truck-mounted mobile pretreatment. A hydrolysate can be obtained onsite and then shipped to a central processing unit for ethanol production, or fermented onsite to produce bioethanol for local use as a fuel. If green pretreatment methods are used, the biomass wastes can be disposed of locally without long-distance hauling. Lignin can be burned as fuel or as a feedstock for producing other useful chemicals using lignin chemistry. Xylose from hemicellulose may be fermented to produce ethanol using special microbes that can utilize it as a carbon energy source. Genetically engineered yeast (Bera et al. 2011), *Zymomonas mobilis* (Yanase et al. 2012) and *Escherichia coli* (Alterthum and Ingram 1989) were found capable of xylose fermentation. Xylose may also be co-fermented with glucose without separation from glucose (Bera et al. 2011).

The localized pretreatment of bulky biomass reduces costs associated with biomass handling, transportation, and storage. Unused by-products can be disposed of to return nitrogen, potassium, and other nutrients to the local soil. Because mobile or on-farm facility may not be large enough to afford waste treatment, green pretreatment methods are preferred.

Ammonia explosion with ammonia recycle is an effective pretreatment method. However, a potential ammonia leak can be very dangerous. The widely used steam explosion method requires a much higher temperature that is better suited for a large facility. Acid hydrolysis is another widely used pretreatment method, but it produces acid wastewater that requires neutralization before disposal. This neutralization process produces gypsum that is difficult to dispose of.

The $SC\text{-}CO_2$ pretreatment method is particularly suited for mobile and on-farm pretreatment because it produces no wastes and its pressure and temperature requirements are tenable (Narayanaswamy et al. 2011). CO_2 can be transported in solid (dry ice) or liquid form to the work site. It can be safely released after use without polluting the environment. Narayanaswamy (2010) presented a flowchart for an $SC\text{-}CO_2$ mobile pretreatment unit for mobile and on-farm biomass pretreatment.

5.7 Conclusion

Several pretreatment methods for lignocellulosic biomass can be considered green because they do not use toxic chemicals and require no special waste treatment. They include mechanical pretreatment, biological pretreatment, H_2O pretreatment (steam explosion and supercritical hot water), IL pretreatment, and $SC\text{-}CO_2$ pretreatment. CO_2 is a nontoxic green solvent. $SC\text{-}CO_2$ pretreatment may utilize some of the CO_2 produced during ethanol fermentation, thus resulting in no extra CO_2 emission into the atmosphere. CO_2 at supercritical state penetrates deeper than other fluids into the lignocellulosic biomass and the CO_2 explosion opens up pores to increase the accessible surface areas for the subsequent enzymatic hydrolysis. This method also benefits from the acidity introduced by the dissolved CO_2 into the water in moist biomass. The pretreatment is very effective for various types of biomass such as hardwood, softwood, corn stover, and switchgrass under certain operating conditions. This chapter discussed various operating parameters used in $SC\text{-}CO_2$ pretreatment of different types of lignocellulosic biomass. It was found that the optimal operating conditions in the literature were different for various biomass types. Compared with H_2O pretreatment, $SC\text{-}CO_2$ pretreatment uses lower temperatures and pressures. The amount of CO_2 to dry biomass mass ratio is much less than those used in AFEX and dilute acid pretreatment methods. This method is particularly suited for mobile biomass pretreatment because the operating conditions are readily met on a truck amounted system and no harmful chemicals are discharged. Research is needed to investigate the possibility of CO_2 recycling without increasing capital and operating costs considerably.

References

Alizadeh H, Teymouri F, Gilbert TI, Dale BE (2005) Pretreatment of switchgrass by ammonia fibere xplosion(AFEX). ApplB iochemB iotechnol124: 1133–1141

Alterthum F, Ingram LO (1989) Efficient ethanol production from glucose, lactose, and xylose by recombinant Escherichia coli. Appl Environ Microbiol 55:1943–1948

Artz W, Warren C, Villota R (1990) Twin-screw extrusion modification of a corn fiber and corn starch extruded blend. J Food Sci 55:746–754

Bera AK, Ho NWY, Khan A, Sedlak M (2011) A genetic overhaul of Saccharomyces cerevisiae 424A (LNH-ST) to improve xylose fermentation. J Ind Microbiol Biotechnol 38:617–626

Buchanan BB, Gruissem W, Jones RL (2000) Biochemistry and molecular biology of plants. Wiley,N ew York

Carrillo F, Lis M, Colom X, López-Mesas M, Valldeperas J (2005) Effect of alkali pretreatment on cellulase hydrolysis of wheat straw: kinetic study. Process Biochem 40:3360–3364

Demirbas A(2007)P rogressa ndr ecentt rendsi nbi ofuels.P rogE nergyC ombustS ci33: 1–18

FelderR M,R ousseauR W(2005)El ementarypr inciplesof c hemicalpr ocesses. Wiley,N ew York

Gao M, Xu F, Li S, Ji X, Chen S, Zhang D (2010) Effect of SC-CO2 pretreatment in increasing rice straw biomass conversion. Biosyst Eng 106:470–475

Garrote G, Dominguez H, Parajo JC (1999) Hydrothermal processing of lignocellulosic materials. Eur J Wood Wood Prod 57:191–202

Hendriks A, Zeeman G (2009) Pretreatments to enhance the digestibility of lignocellulosic biomass.B ioresour Technol100: 10–18

Hu Z, Wen Z (2008) Enhancing enzymatic digestibility of switchgrass by microwave-assisted alkali pretreatment. Biochem EngJ 38: 369–378

Kaar W, Gutierrez C, Kinoshita C (1998) Steam explosion of sugarcane bagasse as a pretreatment for conversion to ethanol. Biomass Bioenergy 14:277–287

Karunanithy C, Muthukumarappan K (2011) Optimization of switchgrass and extruder parameters for enzymatic hydrolysis using response surface methodology. Ind Crops Prod 33:188–199

Keller FA, Hamilton JE, Nguyen QA (2003) Microbial pretreatment of biomass. Appl Biochem Biotechnol105:27 –41

Kim KH, Hong J (2001) Supercritical CO_2 pretreatment of lignocellulose enhances enzymatic cellulose hydrolysis. Bioresour Technol77: 139–144

Kim TH, Lee YY (2005) Pretreatment and fractionation of corn stover by ammonia recycle percolation process. Bioresour Technol96: 2007–2013

Laser M, Schulman D, Allen SG, Lichwa J, Antal MJ, Lynd LR (2002) A comparison of liquid hot water and steam pretreatments of sugar cane bagasse for bioconversion to ethanol. Bioresour Technol81: 33–44

Laureano-Perez L, Teymouri F, Alizadeh H, Dale BE (2005) Understanding factors that limit enzymatic hydrolysis of biomass. In: Twenty-sixth symposium on biotechnology for fuels and chemicals, ABAB symposium series, Davison, Brian H. (Ed.), pp 1081–1099, Springer, Berlin-New York

Li C, Knierim B, Manisseri C, Arora R, Scheller HV, Auer M, Vogel KP, Simmons BA, Singh S (2010) Comparison of dilute acid and ionic liquid pretreatment of switchgrass: biomass recalcitrance, delignification and enzymatic saccharification. Bioresour Technol 101:4900–4906

Lloyd TA, Wyman CE (2005) Combined sugar yields for dilute sulfuric acid pretreatment of corn stover followed by enzymatic hydrolysis of the remaining solids. Bioresour Technol 96:1967–1977

Luterbacher JS, Tester JW, Walker LP (2010) High-solids biphasic CO_2–H_2O pretreatment of lignocellulosic biomass. Biotechnol Bioeng107: 451–460

Luterbacher JS, Tester JW, Walker LP (2012) Two-temperature stage biphasic CO_2–H_2O pretreatment of lignocellulosic biomass at high solid loadings. Biotechnol Bioeng 109:1499–1507

Mamman AS, Lee JM, Kim YC, Hwang IT, Park NJ, Hwang YK, Chang JS, Hwang JS (2008) Furfural: hemicellulose/xylosederivedbi ochemical.B iofuels,B ioprodB iorefin2: 438–454

Meyssami B, Balaban MO, Teixeira AA (1992) Prediction of pH in model systems pressurized with carbon dioxide. Biotechnol Prog 8:149–154

Montane D, Farriol X, Salvadó J, Jollez P, Chornet E (1998) Fractionation of wheat straw by steam-explosion pretreatment and alkali delignification. Cellulose pulp and byproducts from hemicellulosea ndl ignin.J WoodC hem Technol18: 171–191

Mosier N, Wyman C, Dale B, Elander R, Lee Y, Holtzapple M, Ladisch M (2005) Features of promising technologies for pretreatment of lignocellulosic biomass. Bioresour Technol 96:673–686

Muratov G, Seo KW, Kim C (2005) Application of supercritical carbon dioxide to the bioconversion of cotton fibers.J I ndE ngC hem11: 42–46

Narayanaswamy N (2010) Supercritical carbon dioxide pretreatment of various lignocellulosic biomasses. MS Thesis, Ohio University, Athens

Narayanaswamy N, Faik A, Goetz DJ, Gu T (2011) Supercritical carbon dioxide pretreatment of corn stover and switchgrass for lignocellulosic ethanol production. Bioresour Technol 102:6995–7000

Pérez J, Ballesteros I, Ballesteros M, Sáez F, Negro M, Manzanares P (2008) Optimizing liquid hot water pretreatment conditions to enhance sugar recovery from wheat straw for fuel-ethanol production. Fuel 87:3640–3647

Peterson CL, Hustrulid T (1998) Carbon cycle for rapeseed oil biodiesel fuels. Biomass Bioenergy14:91–101

RFA (2011) Building bridges to a more sustainable future: 2011 Ethanol industry outlook. RenewableFue ls Association http://www.ethanolrfa.org/. Accessed28J une2012

Simmons BA, Singh S, Holmes BM, Blanch HW (2010) Ionic liquid pretreatment. Chem Eng Prog106:50–55

Srinivasan N, Ju LK (2010) Pretreatment of guayule biomass using supercritical carbon dioxide-based method. Bioresour Technol101: 9785–9791

Teymouri F, Laureano-Perez L, Alizadeh H, Dale BE (2004) Ammonia fiber explosion treatment of corn stover. Appl Biochem Biotechnol 115:951–963

Teymouri F, Laureano-Perez L, Alizadeh H, Dale BE (2005) Optimization of the ammonia fiber explosion (AFEX) treatment parameters for enzymatic hydrolysis of corn stover. Bioresour Technol96: 2014–2018

Wan C, Li Y (2010) Microbial pretreatment of corn stover with *Ceriporiopsis subvermispora* for enzymatic hydrolysis and ethanol production. Bioresour Technol101: 6398–6403

Wang G, Lee JW, Zhu J, Jeffries TW (2011) Dilute acid pretreatment of corncob for efficient sugar production. Appl Biochem Biotechnol 163:658–668

Wyman CE, Dale BE, Elander RT, Holtzapple M, Ladisch MR, Lee Y (2005) Comparative sugar recovery data from laboratory scale application of leading pretreatment technologies to corn stover.B ioresour Technol96: 2026–2032

Xu L, Tschirner U (2012) Peracetic acid pretreatment of alfalfa stem and aspen biomass. BioResources7:203–216

Yanase H, Miyawaki H, Sakurai M, Kawakami A, Matsumoto M, Haga K, Kojima M, Okamoto K (2012) Ethanol production from wood hydrolysate using genetically engineered Zymomonasmobi lis. ApplM icrobiolB iotechnol94: 1667–1678

Yang B, Wyman CE (2008) Pretreatment: the key to unlocking low-cost cellulosic ethanol. Biofuels,B ioprodB iorefin2: 26–40

Zheng Y, Lin HM, Tsao GT (1998) Pretreatment for cellulose hydrolysis by carbon dioxide explosion. Biotechnol Prog 14:890–896

Zheng Y, Lin HM, Wen J, Cao N, Yu X, Tsao GT (1995) Supercritical carbon dioxide explosion as a pretreatment for cellulose hydrolysis. Biotechnol Lett 17:845–850

Chapter 6
Pretreatment of Lignocellulosic Biomass Using Green Ionic Liquids

Jian Luo, Meiqiang Cai and Tingyue Gu

Abstract Bioenergy is a critical part of renewable energy solution to today's energy crisis that threatens world economic growth. Corn ethanol has been growing rapidly in the past few years. Policy-makers and researchers alike are becoming aware that corn ethanol has some serious drawbacks. It adversely impacts food prices and is harsh on soil fertility. Lignocellulosic ethanol on the other hand uses abundant lignocellulosic biomass. Various types of lignocellulosic biomass are agricultural wastes or can be grown as energy crops on poor lands that are otherwise vacant. However, lignocellulosic biomass is notoriously recalcitrant by nature's design. Enzyme hydrolysis yield for glucose from lignocellulosic is very low without proper pretreatment. Quite a few pretreatment methods have been reported in the literature. For mobile and on-farm biomass processing, a pretreatment method that uses no chemicals or green chemicals without the need for waste treatment is preferred. In recent years, research on using ionic liquids as green solvents for pretreatment of lignocellulosic biomass has exploded. Hundreds of papers have been published in the literature in the past few years alone. Some ionic liquids such as [Amim]Cl and [C$_2$mim]OAc have been proven highly effective in the dissolution of cellulose, lignin, and hemicellulose in different types of lignocellulosic biomass including corn stover, switchgrass, rice straw, and various hard and softwoods. This simple pretreatment method has been proven highly effective for improving sugar yields in the enzyme hydrolysis of the recovered biomass after pretreatment. Various methods have been developed for ionic liquid recycling after pretreatment. Although costly, ionic liquids hold great potential for green pretreatment of biomass as the technology improves. This chapter investigated the mechanisms and various parameters in ionic liquid pretreatment of various types of lignocellulosic biomass.

J. Luo
National Key Lab of Biochemical Engineering, Institute of Process Engineering,
Beijing 100190, China

M. Cai
College of Environmental Science and Engineering, Zhejiang Gongshang University,
Hangzhou 310035, China

T. Gu (✉)
Department of Chemical and Biomolecular Engineering, Ohio University,
Athens, OH 45701, USA
e-mail: gu@ohio.edu

T. Gu (ed.), *Green Biomass Pretreatment for Biofuels Production*,
SpringerBriefs in Green Chemistry for Sustainability,
DOI: 10.1007/978-94-007-6052-3_6, © The Author(s) 2013

Keywords Ionic   liquid • Biomass   pretreatment • Biorefinery • Cellulose • Hemicellulose • Lignin

6.1 Introduction

An integrated approach using different forms of renewable energy such as wind, solar, and biomass together with nuclear energy is necessary to supplement the shortfall caused by dwindling fossil fuel (especially petroleum) reserves. Bioethanol has seen a tremendous growth in the last few years. However, this growth cannot be sustained. Due to increased farmland use for corn ethanol, food prices have been increasing. In fact, 25-gallon of corn ethanol filled into the gas tank of a large sports utility vehicle consumes up to 450 pounds of corn. This much corn has sufficient calories to feed one person for a whole year (Runge and Senauer 2007).

Various types of lignocellulosic biomass such as corn stover, rice straw, wood, are abundantly available as agricultural wastes. They may be used as renewable feedstocks for a biorefinery that produces biofuels and chemicals. Some energy crops such as switchgrass may be planted on poor lands that are vacant. Unfortunately, lignocellulosic biomass is very recalcitrant. Without proper pretreatment, it is difficult to release fermentable sugars such as glucose using enzyme hydrolysis. The pretreatment step is often the most expensive part of a lignocellulosic ethanol process. It is critical to improve this step in order to make lignocellulosic ethanol economically competitive.

Chapter 1 in this book discussed cell wall structures and chemical components. Typical lignocellulosic biomass contain lignin (15–25 % w/w), hemicellulose (23–32 %), and cellulose (38–50 %) (Mamman et al. 2008). Cellulose is a glucan biopolymer containing glucopyranose subunits with a molecular formula of $(C_6H_{10}O_5)_n$. Upon hydrolysis, each subunit gains one H_2O molecule. Thus, the maximum theoretical glucose yield for 100 g pure cellulose is 111 g. The glucopyranose subunits in the cellulose are linked by β-1,4-glycosidic bonds that are highly resistant to hydrolysis. Apart from this, enzyme hydrolysis is greatly hindered by the crystallinity of cellulose and the protective sheath of lignin and hemicellulose that wrap around cellulose (Laureano-Perez et al. 2005). An effective pretreatment method can weaken all these hindrances and exposes cellulose to cellulase enzymes for effective hydrolysis. Alizadeh et al. (2005) reported that only less than 20 % glucose is released from lignocellulosic biomass without pretreatment while the yield can be as high as 90 % with proper pretreatment. However, pretreatment is often expensive. It constitutes 20 % of the overall process cost for lignocellulosic ethanol (Yang and Wyman 2008). Thus, it is imperative to improve pretreatment methods.

Many types of lignocellulosic biomass such as corn stover and switchgrass are bulky and costly to transport on a mass basis to a central processing facility. Thus, it is desirable in some situations to use mobile (tactical) or on-farm processing to eliminate the need for biomass transportation. It is possible to integrate pretreatment, enzyme hydrolysis, fermentation, and ethanol fermentation and separation to produce fuel ethanol on farms where certain types of lignocellulosic biomass are abundant.

It can be economical even at relatively small scales to supply such fuels when the local supplies are unusually expensive, such as in some military forward operating bases.

Many pretreatment methods have been proposed for lignocellulosic biomass. They include mechanical pretreatment, biological pretreatment, acid and alkaline pretreatment, steam explosion pretreatment, (supercritical) hot water pretreatment, ammonia pretreatment, supercritical CO_2 explosion pretreatment and ionic liquid pretreatment (Zheng et al. 1995; Teymouri et al. 2004; Alizadeh et al. 2005; Lloyd and Wyman 2005; Mosier et al. 2005; Kim and Lee 2007; Hendriks and Zeeman 2009; Wan and Li 2010; Karunanithy and Muthukumarappan 2011; Narayanaswamy et al. 2011; Wang et al. 2012). Among the chemical pretreatment methods, CO_2 and ionic liquids are considered green treatment methods because CO_2 and ionic liquids are green solvents. Chapter 5 in this book discussed CO_2 pretreatment in depth already.

The three major components in lignocellulosic biomass (lignin, hemicellulose, and cellulose) are biopolymers in an organized network that are considered insoluble because they cannot be dissolved in common solvents. If a special solvent that is able to dissolve any of the three components, it would weaken the biomass structure and makes the biomass less recalcitrant for enzyme hydrolysis. Ionic liquids are such solvents that can be used in the pretreatment step to achieve some or all of the following objectives that help to reduce the biomass recalcitrance: (1) Amorphization of cellulose, (2) delignification, and (3) deacetylation of hemicellulose.

Ionic liquids are a special group of organic salts that can exist in liquid form at relatively low temperatures (less than 100 °C) and some can even exist as liquids at room temperature (Cull et al. 2000; Marsh et al. 2004; Zhu et al. 2006). They are usually molten salts or oxides (Marsh et al. 2004). Thus, they have much higher viscosities (similar to those for oils) than conventional organic solvents and most of them are heavier than water (Marsh et al. 2004). Ionic liquids are special solvents that can dissolve chemical compounds that are otherwise considered insoluble in conventional solvents. Figure 6.1 shows the complete dissolution of tiny wood chips with the ionic liquid [C_2mim]OAc.

Typically in ionic liquids, the cations are organic while the anions may be inorganic or organic (Marsh et al. 2004; Wang et al. 2012). Some common cations and anions for ionic liquids used for biomass dissolution are listed in Table 6.1. Ionic liquids have found applications in many different areas, such as analytical chemistry (Baker et al. 2005; Han and Armstrong 2007), chemical reactions including biotransformation reactions (Cull et al. 2000; Sun et al. 2011), extractive separations of metal ions, proteins and other organic molecules (Huddleston and Rogers 1998; Han and Armstrong 2007), and lignocellulosic biomass pretreatment and processing (Zhu et al. 2006; Wang et al. 2012). Chemical compounds extracted or dissolved in ionic liquids may be recovered by using an antisolvent such as water, ethanol, acetone, or even supercritical CO_2 (Blanchard and Brennecke 2001; Zhu et al. 2006).

Graenacher (1934) was the first to file a patent on the use of an ionic liquid for cellulose dissolution. He claimed to have used molten N-ethylpyridinium chloride (an ionic liquid) in the presence of nitrogen-containing bases to dissolve cellulose without derivation. Not enough attention was paid to this kind of application because there was no great push for cellulosic ethanol and green chemistry at the time. Although some of the claims in the patent were found to be inaccurate by Sun et al. (2011), Graenacher was

Fig. 6.1 Dissolution of tiny wood chips from common beech in an ionic liquid (*left*: blank [C₂mim]OAc; *middle*: wood chips from common beech; *right*: a homogeneous ionic liquid solution after dissolution) (Reprinted from Bioresource Technology, 100/9, High-throughput screening for ionic liquids dissolving (ligno-)cellulose, Michael Zavrel, Daniela Bross, Matthias Funke, Jochen Büchs, Antje C. Spiess, 8, Copyright 2009, with permission from Elsevier)

still considered the first person who brought up the use of ionic liquids for cellulose dissolution. It was Swatloski et al. (2002) who ignited this research area by publishing a widely cited study in 2002 on the use of ionic liquids for dissolution of pulp cellulose. They also showed that cellulose dissolved in ionic liquids could be precipitated using water. There has been an explosion in the number of publications dealing with ionic liquid pretreatment in the past few years. Dozens of ionic liquids are found to dissolve cellulose (Shill et al. 2011; Wang et al. 2012). The dissolved cellulose can be recovered and hydrolyzed using enzymes with much higher glucose yields. Research in this area was also expanded to raw biomass containing lignin and hemicellulose in addition to cellulose that was used as a model biomass for early investigations in its essentially pure form (e.g., Avicel, microcrystalline cellulose, and pulp cellulose). Many ionic liquids are found to dissolve lignin and hemicellulose as well (Lee et al. 2009; Zavrel et al. 2009; Fu et al. 2010; Casas et al. 2012; Lynam et al. 2012; Xin et al. 2012). By removing lignin and hemicellulose, enzyme hydrolysis of cellulose can also be improved greatly.

This chapter investigated the mechanisms for ionic liquid pretreatment and discusses various key operating parameters involved in the ionic liquid pretreatment of different types of lignocellulosic biomass. An emphasis is placed on how ionic liquid could be used for green mobile or on-farm pretreatment.

6.2 Mechanisms for Ionic Liquid Dissolution of Biomass

Swatloski et al. (2002) proposed that the high Cl^- concentration and activity in an ionic liquid such as [C₄mim]Cl can effectively break the extensive network of intra- and intermolecular hydrogen bonds in cellulose, thus allowing cellulose

Table6.1 Some common cations and anions in ionic liquids used in biomass dissolution

Type	Formula	Structure	Fullna me
Cations	$[Amim]^+$		1-allyl-3-methylimidazolium
	$[C_1mim]^+$,$[Mmim]^+$		1,3-dimethylimidazolium
	$[C_2mim]^+$,$[Emim]^+$		1-ethyl-3-methylimidazolium
	$[C_3mim]^+$,		1-propyl-3-methylimidazolium
	$[C_4mim]^+$,$[Bmim]^+$		1-butyl-3-methylimidazolium
	$[C_6mim]^+$, $[Hmim]^+$		1-hexyl-3-methylimidazolium
	$[C_8mim]^+$		1-octyl-3-methylimidazolium
	$[Cholinium]^+$		2-hydroxy-N,N,N-trimethyl-ethanaminium
Anions	Cl^-		Chloride
	CH_3COO^- (OAc^-, Ac^-)		Acetate
	$HCOO^-$		Formate
	$(C_2H_5O)_2(PO_2)^-$ (DEP)		Diethyl phosphate
	Gly^-		Glycine
	Lys^-		Lysine

dissolution in the ionic liquid. For example, decreased Cl^- concentrations (e.g., when using $[C_6mim]Cl$ and $[C_8mim]Cl$ instead of using $[C_4mim]Cl$) led to lower cellulose solubility in the ionic liquids. They also observed a significant impairment of cellulose solubility by adding as little as 1 wt% water in the ionic liquid, suggesting that competitive hydrogen bonding of water with cellulose was likely to blame for the impairment. This effect also means that water can be used to precipitate cellulose from ionic liquids. The NMR data presented by Remsing et al. (2006)

confirmed that the solvation of cellulose by [C$_4$mim]Cl had a 1:1 molar ratio for hydrogen-bonding between Cl$^-$ anions in the ionic liquid and the hydroxyl hydrogen atoms in cellulose polymer. Figure 6.2 shows a schematic presentation of solvation mechanism of cellulose using [C$_4$mim]Cl as an example.

Some other anions that are good hydrogen bond acceptors such as OAc$^-$, HCOO$^-$ and (C$_2$H$_5$O)$_2$(PO$_2$)$^-$ have also been found effective for cellulose dissolution in ionic liquids. Acetate anion (OAc$^-$) is considered more effective than chloride anion (Wang et al. 2012). Zhang et al. (2010) presented NMR data that suggested that acetate anions in [C$_2$mim]OAc formed hydrogen bonds with the hydroxyl hydrogen atoms in cellobiose that is the repeating subunit in cellulose. They found that acetylation of the hydroxyl groups in cellobiose led to weakened solvation of cellobiose by [C$_2$mim]OAc. NMR and solvatochromic studies of microcrystalline cellulose dissolution in ionic liquids containing [C$_4$mim]$^+$ cation paired with Brønsted basic anions [CH$_3$COO]$^-$, [HSCH$_2$COO]$^-$, [HCOO]$^-$, [(C$_6$H$_5$)COO]$^-$, [H$_2$NCH$_2$COO]$^-$, [HOCH$_2$COO]$^-$, [CH$_3$CHOHCOO]$^-$ and [N(CN)$_2$]$^-$ carried out by Xu et al. (2010) showed that cellulose solubility increased nearly linearly with the anion's hydrogen-bonding accepting ability. They also found that pretreatment temperature had a major impact on cellulose solubility in the ionic liquids. Molecular dynamics simulation supported the NMR observation of hydrogen bonding between cellobiose's hydroxyl hydrogen atoms with Cl$^-$ anions in [C$_4$mim]Cl (Novoselov et al. 2007), and between cellulose's hydroxyl hydrogen atoms and OAc$^-$ anions in [C$_2$mim]OAc (Liu et al. 2010). Ionic liquids containing anions that are low basicity anions such as BF$_4^-$ or PF$_6^-$ were found ineffective for cellulose dissolution (Swatloski et al. 2002). It should be pointed out that some ionic liquids such as those containing [PF$_6$] and [BF$_4$] anions also have a tendency to decompose (Marsh et al. 2004).

Experimental results (Wang et al. 2012) and molecular dynamics simulation (Liu et al. 2010) seem to suggest that cations in ionic liquids also play a role in solvation of cellulose. It was found that [C$_2$mim]Cl, [C$_4$mim]Cl and [C$_6$mim]Cl had good solubility for cellulose while [C$_3$mim]Cl and [C$_5$mim]Cl had almost no solubility and a low solubility, respectively. Wang et al. (2012) speculated that such disparities in cellulose solubility might be because cations played a role in solvating and dispersing the anionic hydrogen bonded cellulose•Cl$_n$ moieties. Based on molecular dynamics simulation results, Liu et al. (2010) suggested that hydrophobic interactions between imidazolium cations in ionic liquids and glucopyranose subunits in the cellulose polymer play an important role in cellulose dissolution. The roles of cations are still somewhat controversial. More investigative work is needed (Sun et al. 2011; Wang et al. 2012).

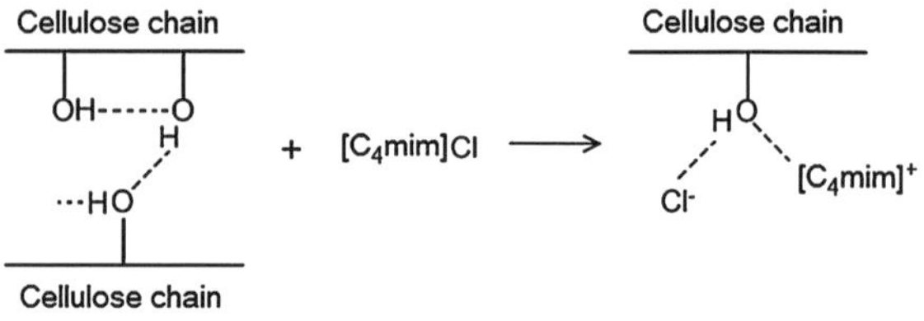

Fig. 6.2 Possible mechanism for cellulose dissolution in [C$_4$mim]Cl (A daptedf rom Fenga ndC hen2008)

When any of the three components in lignocellulosic biomass (cellulose, hemicellulose, and lignin) is dissolved in an ionic liquid, the extensive network within plant cell wall is disrupted. This reduces recalcitrance of the biomass. The cellulose regenerated after the pretreatment tends to be more amorphous in its macrostructure and thus is easier for enzyme hydrolysis (Wang et al. 2012). Cheng et al. (2011) studied the cellulose crystallinity transition and surface morphology of Avicel, switchgrass, pine wood, and eucalyptus wood after [C_2mim]OAc pretreatment at 120 and 160 °C for 1–12 h. The pretreatment resulted in the loss of native cellulose crystalline I structure. Avicel was easier to transform into the cellulose II structure than switchgrass and eucalyptus, while the cellulose II structure was not detected for pine after 12 h at 160 °C. Singh et al. (2009) visualized switchgrass dissolution in [C_2mim]OAc through confocal fluorescence images and found that in 2 h the ionic liquid completely broke down the organized cell wall structure (Fig. 6.3). Their analysis of SEM images of switchgrass before and after the pretreatment indicated that there was no lignin accumulation in pretreated and regenerated cellulose (Fig. 6.4).

Ionic liquid pretreatment of biomass has gone far beyond the initial intention of just dissolving cellulose. Lignin and hemicellulose can also be selectively dissolved in some ionic liquids, thus providing more options for pretreatment of lignocellulosic biomass. Liu et al. (2012) synthesized a series of ionic liquids containing cholinium cation paired with different amino acids as the anion. They found that [Cholinium]Gly pretreatment of rice straw improved enzyme hydrolysis yield of glucose by several folds due to lignin removal by the ionic liquid, which was found to have a high lignin solubility of 220 mg/g and a low cellulose solubility of less than 5 mg/g at 90 °C. Lee et al. (2009) found that when 40 % of the lignin in wood flour was dissolved in [C_2mim]OAc, the crystallinity of the cellulose component was much reduced without cellulose dissolution in the ionic liquid. This led to greater than 90 % cellulose hydrolysis by *Trichoderma viride* cellulase. They noticed that when a type of lignin had a similar Hildebrand solubility parameter as that for the ionic liquid used for its dissolution, the lignin solubility was high because solubility of polymers in solvents could be predicted by the Hildebrand solubility parameter. For [C_1mim][$MeSO_4$] and [C_4mim][CF_3SO_3], lignin solubility greater than 0.5 kg/kg ionic liquid were observed by Lee et al. (2009). These ionic liquids have low solubility for cellulose and hemicellulose and the regenerated lignin after dissolution was chemically unaltered. Apart from using cellulose to yield glucose for bioethanol production, lignin can be used as a renewable chemical feedstock for many other applications (Lee et al. 2009). Tan et al. (2009) successfully extracted lignin from bagasse using an ionic liquid consisting of [C_2mim]$^+$ cation and a mixture of alkylbenzenesulfonates and xylenesulfonate anions. The dissolved lignin was subsequently separated from dissolved cellulose using aqueous sodium hydroxide. Fort et al. (2007) demonstrated that [C_4mim]Cl could dissolve woods with different hardness for further processing. Lignin extraction from biomass using an ionic liquid containing chloride anion requires a virtual absence of water moisture in biomass and the ionic liquid, and an inert atmosphere (Tan and MacFarlane 2010; Shill et al. 2011). It was suggested that lignin dissolution was facilitated by the π–π interactions between the cation and the lignin (Shill et al. 2011).

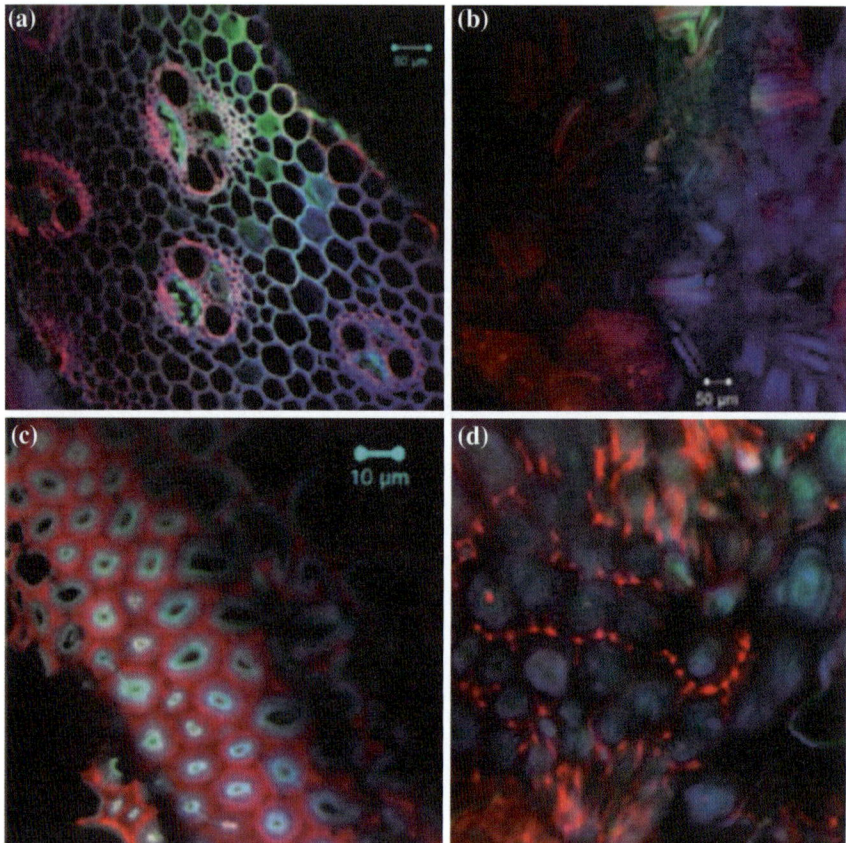

Fig. 6.3 Confocal fluorescence images of switchgrass stem section before and after [C₂mim]OAc pretreatment: **a** before pretreatment, **b** after 20 min pretreatment, **c** after 50 min pretreatment, and **d** after 2 h pretreatment. (Reprinted from Biotechnology and Bioengineering, Visualization of biomass solubilization and cellulose regeneration during ionic liquid pretreatment of switchgrass, Seema Singh, Blake A. Simmons, Kenneth P. Vogel,104:68–75, Copyright 2009, with permission from John Wiley and Sons)

Lynam et al. (2012) found that pretreatment of rice hulls at 110 °C for 8 h in [C₂mim]OAc dissolved 100 % of lignin and up to 29 % hemicellulose, but it was ineffective in dissolving cellulose. However, [Amim]Cl under the same conditions dissolved 33 % cellulose and 75 % hemicellulose. For [C₆mim]Cl, the numbers were 23 and 70 %, respectively. Both [Amim]Cl and [C₆mim]Cl were ineffective in dissolving lignin. The impact of ionic liquid pretreatment on hemicellulose was studied using NMR by Çetinkol et al. (2010), who observed deacetylation of hemicellulose when *Eucalyptus globules* biomass was dissolved in [C₂mim]OAc, leading to enhanced release of xylose during saccharification. They also found that the syringyl to guaiacyl ratio in lignin increased after the pretreatment and there was possible acetylation of β-aryl ether linkages in lignin.

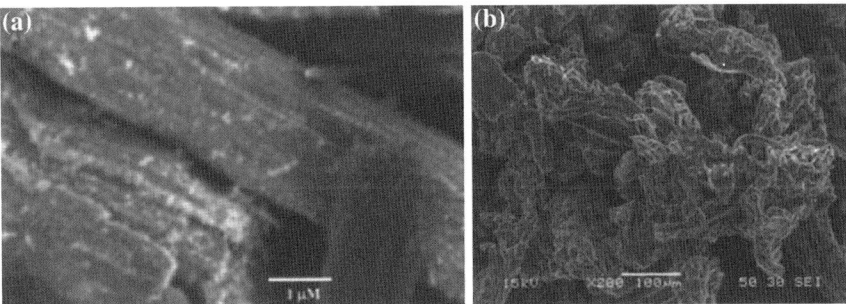

Fig. 6.4 SEM images of untreated (**a**), and ionic liquid treated and recovered switchgrass fibers (**b**). (Reprinted from Biotechnology and Bioengineering, Visualization of biomass solubilization and cellulose regeneration during ionic liquid pretreatment of switchgrass, Seema Singh, Blake A. Simmons, Kenneth P. Vogel, 104:68–75, Copyright 2009, with permission from John Wiley and Sons)

Ionic liquid pretreatment is comparable to other mainstream pretreatment methods in sugar yields, but it requires no harsh operating conditions such as high temperature and high pressure. The maximum sugar conversion could reach 90.0 % (w/w) based on regenerated biomass from cassava pulp residue treated with [C$_2$mim]OAc at 120 °C for 24 h (Weerachanchai et al. 2012). Xu et al. (2012) reported that after [C$_2$mim]OAc pretreatment of corn stover at a relatively low pretreatment temperature of 70 °C for 24 h, they achieved glucose, xylose and total sugar yields of 84.9, 64.8, and 78.0 %, respectively, which are comparable to the yields of 82.0, 72.2, and 78.4 % obtained using the Ammonia Fiber Expansion (AFEX) pretreatment method by Li et al. (2011a, b) as shown in Fig. 6.5. These were achieved when corn stover was not completely dissolved in the ionic liquid. Another group of researchers compared dilute acid pretreatment with ionic liquid pretreatment of switchgrass experimentally (Li et al. 2010). For ionic liquid pretreatment, they used [C$_2$mim][OAc] at 160 °C for 3 h. For dilute acid pretreatment, the switchgrass sample was presoaked at room temperature in 1.2 % (w/w) sulfuric acid for at least 4 h before the reactor temperature was raised to 160 °C for 20 min. Figure 6.6 shows that yields of different sugars including glucose and xylose after enzymatic saccharification were slightly lower for ionic liquid pretreatment.

Many factors impact ionic liquid pretreatment of biomass. They include ionic liquid type, ionic liquid to biomass mass ratio, pretreatment time and temperature, and water content, etc. These factors are discussed in the sections below.

6.3 Selection of Ionic Liquids

Clearly, there are several possible basic strategies for ionic liquid pretreatment of lignocellulosic biomass: (1) Dissolve cellulose in processed cellulose material such as pulp cellulose in an ionic liquid, (2) remove lignin (and hemicellulose) by

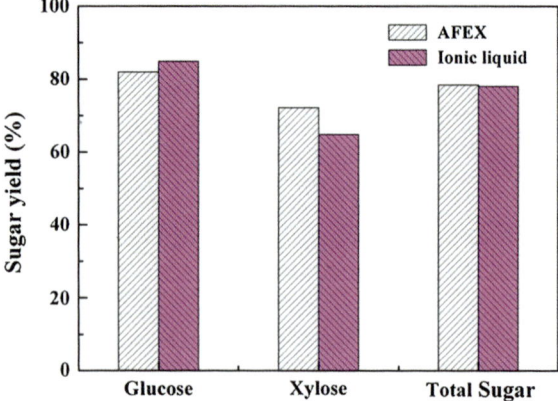

Fig.6.5 Enzyme hydrolysis sugar yields from corn stover with pretreatment using [C₂mim]OAc at 70 °C for 24 h compared with those with pretreatment using AFEX (Figure plotted with data from Xu et al. 2012)

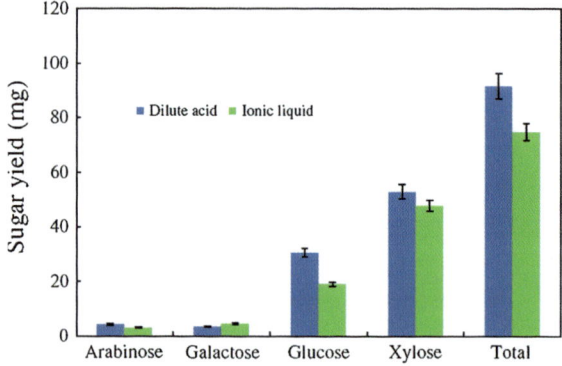

Fig. 6.6 Various enzyme hydrolysis sugar yields from switchgrass pretreated in an ionic liquid compared with those with pretreatment using dilute acid. (Reprinted from Bioresource Technology, 101/13, Comparison of dilute acid and ionic liquid pretreatment of switchgrass: Biomass recalcitrance, delignification, and enzymatic saccharification, Chenlin Li, Bernhard Knierim, Chithra Manisseri, Rohit Arora, Henrik V. Scheller, Manfred Auer, Kenneth P. Vogel, Blake A. Simmons, Seema Singh, 4900–4906, Copyright 2010, with permission from Elsevier)

dissolving the lignocellulosic biomass in an ionic liquid while leaving cellulose behind, and (3) dissolve all the biomass components and then selectively recover the needed component(s). Table 6.2 shows some ionic liquids that have been used for lignocellulosic biomass pretreatment in the literature.

An effective ionic liquid for cellulose dissolution in biomass pretreatment should satisfy the following three criteria: (1) Its anion is a good hydrogen bond acceptor; (2) its cation should be a moderate hydrogen bond donor, and (3) its cation should not be too bulky (Fukaya et al. 2008; Mäki-Arvela et al. 2010; Feng and Chen 2008).

From the perspective of process engineering, a good ionic liquid for cellulose dissolution should have high cellulose solubility, low melting point, low viscosity, good stability, and low toxicity (Sun et al. 2011). Experimental data suggest that with the same cation, the decreasing order for cellulose solubility for different anions is: $(CH_3CH_2O)_2PO_2^- \approx OAc^- > SHCH_2COO^- > HCOO^- > Cl^- > Br^- \approx SCN^-$ (Swatloski et al. 2002, Klemm et al. 2005). The most effective cations for cellulose dissolution are found to be those based on the methylimidazolium and methylpyridinium cores, and contain allyl-, methyl-, ethyl-, or butyl- side chains (Wang et al. 2012). Some ionic liquids can be used to remove lignin from biomass while leaving behind cellulose. For example, $[C_2mim]OAc$ removes lignin from triticale (a hybrid of wheat and rye) straw efficiently at 70–150 °C in 1.5 h with some hemicellulose removal, but little removal of cellulose (Fu et al. 2010).

The cost of ionic liquid is another important factor in the selection of an ionic liquid for pretreatment. Table 6.3 shows published research chemical prices at www.sigmaaldrich.com in August 2012. The prices may only be used to compare relative prices because large-scale prices will be much lower. Some ionic liquids are about

Table 6.2 Some ionic liquids used in lignocellulosic biomass pretreatment

Ionic Liquid	Biomass	Dissolved components	References
[Amim]Cl	Southern pine TMP	All[a]	Kilpeläinen et al. (2007)
	Avicel cellulose	All	Zhao et al. (2009)
	Corn stalk	All	Li et al. (2008)
	Rice hull	Hemicellulose and some cellulose	Lynam et al. (2012)
[C₂mim]Cl	Avicel	Cellulose	Erdmenger et al. (2007)
[C₄mim]Cl	Pine, poplar, oak, and eucalyptus	All	Fort et al. (2007)
	Pulp cellulose	Cellulose	Swatloski et al. (2002)
	Bagasse	All	Lan et al. (2011)
	Corn stalk and rice straw	All	Li et al. (2008)
[C₆mim]Cl	Corn stalk	All	Li et al. (2008)
	Rice hull	Hemicellulose and some cellulose	Lynam et al. (2012)
[C₂mim]OAc	Softwood and hardwood	All	Sun et al. (2009)
	Triticale straw	Lignin (and a little hemicellulose)	Fu et al. (2010)
	Bagasse and southern yellow pine	All	Li et al. (2011b)
[C₄mim]OAc	Microcrystalline cellulose	Cellulose	Zhao et al. (2008)
	Poplar sawdust	All	Vo et al. (2011)
[C₄mim]HCOO	Avicel cellulose	Cellulose	Zhao et al. (2008)
[C₂mim]DEP	Cassava pulp residue, rice straw	All	Weerachanchai et al. (2012)
[Cholinium]Gly	Rice straw	All	Hou et al. (2012)
[Cholinium]Lys	Rice straw	All	Hou et al. (2012)

[a] Cellulose, hemicellulose, and lignin can all be dissolved

Table6.3 Prices for some ionic liquids used in lignocellulosic biomass pretreatment

IonicLi quid	Prices(US$/g)	Comment
[Amim]Cl	19.14	Basedo n5gpur chase
[Amim]Br	25.60	Basedo n5gpur chase
[C_2mim]Cl	31.80	Basedon5gpur chase
[C_4mim]Cl	11.86	Basedon5gpur chase
[C_2mim]OAc	7.04	Basedon5 gpur chase
[C_4mim]OAc	2.70	Basedon100 gpur chase
[Cholinium]Cl	18.58	Basedo n5gpur chase

5–20 times more expensive than conventional solvents for laboratory applications (Plechkova and Seddon 2008). This is because their synthesis and purification costs are much higher. As ionic liquid applications are expanding, their costs will certainly be more affordable in the future. This may shift biomass loading that is also a key parameter in the selection of ionic fluids for pretreatment (Sect. 6.4).

6.4 BiomassL oading

Biomass loading largely depends on the biomass solubility in the particular ionic liquid at the operating temperature and the biomass dissolution rate that influences pretreatment time. Because the time requirement can be excessively long, thermo-dynamic solubility limit may not be reached during the given pretreatment time. Thus, the ionic liquid to biomass mass ratio should be substantially higher than the solubility limit. A smaller biomass particle size typically means faster biomass dissolution rate. Microwave and ultrasound can also be used to accelerate biomass dissolution (Ha et al. 2011; Wang et al. 2012).

Table 6.4 shows that researchers typically use a nominal biomass loading ratio of 20/1–30/1 (g ionic liquid/g biomass) in ionic liquids pretreatment of various types of biomass under different temperature and time conditions. In comparison, a ratio of 10/1 (g liquid/g biomass) is common for aqueous ammonia, acid hydrolysis, supercritical CO_2, and hydrothermal pretreatment methods (Narayanaswamy 2010; Garrote et al. 1999). It should be noted that if the biomass concentration reaches 10 % (w/w), the ionic liquid solution is almost always very viscous, which makes it difficult to process (Laus et al. 2005). Thus, viscosity could be a limiting factor rather than biomass solubility in a practical pretreatment process.

6.5 Pretreatment Temperature

Apart from selecting a suitable ionic liquid for desired biomass pretreatment out-come, pretreatment temperature is the most important operating parameter (Schultz et al. 1983; Yu et al. 2010). A higher temperature typically achieves higher biomass solubility and shortens the pretreatment time. It reduces the biomass recalcitrance better. However, it also means higher energy input. Compared with steam explosion

Table6.4 Selected data on ionic liquid (IL) pretreatment conditions and biomass dissolution

Biomass	IL	T(°C)	t(h)	IL/bi omass (g/g)	Biomass dissolved (wt%)	References
Southernpine powder	[Amim]Cl	80	8	NM[a]	8	Kilpeläinen et al. (2007)
Woodfl our	[C$_2$mim]OAc	90	1.5	20/1	17	Leee ta l. (2009)
Rices traw	[Cholinium]Gly	90	24	20/1	48.4	Houe ta l. (2012)
Triticales traw	[C$_4$mim]Cl	90	24	20/1	23.1	Fue ta l. (2010)
Southernye llow pine	[C$_2$mim]OAc	110	16	20/1	98.2	Sune ta l. (2009)
Southernye llow pine	[C$_4$mim]Cl	110	16	20/1	52.6	Sune ta l. (2009)
Norways pruce sawdust	[C$_4$mim]Cl	110	8	NM[a]	8	Kilpeläinene ta l. (2007)
Southernpine TMP	[Amim]Cl	130	8	NM[a]	5	Kilpeläinen et al. (2007)
Sugarcane bagasse	[Amim]Cl	140	1	32/1	40.4	Zhue ta l. (2012)
Triticales traw	[C$_2$mim]OAc	150	1.5	20/1	48.8	Fue ta l. (2010)
Corns tover	[C$_2$mim]OAc	160	3	32/1	53.3	Lie ta l. (2011a,b)
Switchgrass	[C$_2$mim]OAc	160	3	32/1	49.3	Lie ta l. (2010)
Wheats traw	[C$_2$mim]OAc	162	4.5	20/1	57.6	Fua ndM azza (2011)
Cassavapul p	[C$_2$mim]DEP	180	24	20/1[b]	88	Weerachanchai et al. (2012)

[a] NM: not mentioned
[b] This special case used 20 ml ionic liquid/1 g biomass

and hot water pretreatment methods, the required pretreatment temperature for ionic liquid pretreatment is relatively mild with a typical temperature range of 80–180 °C (Table 6.4). This is comparable to the typical temperature range used in supercritical CO_2 explosion pretreatment discussed in Chap. 5, but ionic liquid pretreatment does not require pressurization, which is needed by supercritical CO_2. Compared with the typical 140–250 °C temperature range used in steam explosion pretreatment and hot water pretreatment (Schultz et al. 1983; Kaar et al. 1998; Kim and Hong 2001; Yu et al. 2010), ionic liquid has a major advantage. Typically, the upper-end temperature in the temperature ranges is required for the desired pretreatment outcome. Thus, ionic liquid pretreatment enjoys a much lower optimal temperature. This also means that pyrolysis side reaction is easier to avoid when using ionic liquid pretreatment compared with steam or hot water pretreatment.

Fu et al. (2010) treated triticale straw for 1.5 h in [C$_2$mim]OAc at temperatures ranging from 70–150 °C, and then conducted enzyme hydrolysis of residual biomass using *Trichoderma reesei* cellulase at 50 °C. They found that [C$_2$mim]OAc removed little cellulose, but removed lignin efficiently from triticale straw and this reduced the recalcitrance of the remaining cellulose. Figure 6.7 shows that the pretreatment improved cellulose hydrolysis sugar yield greatly and the pretreatment temperature had a strong effect. Higher temperature resulted in better lignin removal and cellulose hydrolysis yield. Complete cellulose hydrolysis (with a hydrolysis time of 11 h) required a pretreatment temperature of 150 °C with a pretreatmenttime of1.5h.

Li et al. (2009) pretreated wheat straw with 1-ethyl-3-methylimidazolium diethyl phosphate ([C$_2$mim]DEP) for 1 h at various temperatures from 25 to 150 °C. Cellulose was precipitated by adding water to the ionic liquid after biomass dissolution. The regenerated cellulose was then hydrolyzed using cellulase. Their data in Fig. 6.8 suggest that 70 °C pretreatment temperature showed no improvement over 25 °C, while 100 °C had a big improvement in reducing sugar yield. Further increasing the pretreatment temperature until 130 °C continued to improve the yield, but increasing the pretreatment temperature from 130 to 150 °C had only a small improvement, indicating that 130 °C could be used as the optimal pretreatment temperature. For lignin removal from rice straw using [Cholinium]Lys, Hou et al. (2012) found that 90 °C pretreatment for 24 h provided the highest glucose yield of 87.7 % after hydrolysis, while 110 and 130 °C actually reduced the yield to 77.1 and 77.2 %, respectively, due to prolonged exposure to the high temperatures. Their data also indicated that 70 °C provided the highest xylose yield of 38.7 %. These relatively low optimal temperatures were partially the result of a rather long pretreatment time of 24 h.

Pezoa et al. (2010) pretreated corn stover with [C$_2$mim]Cl at different temperatures for 60 min. The ionic liquid weakened the biomass by removing lignin while dissolving less than 10 % cellulose. They measured the various sugars released

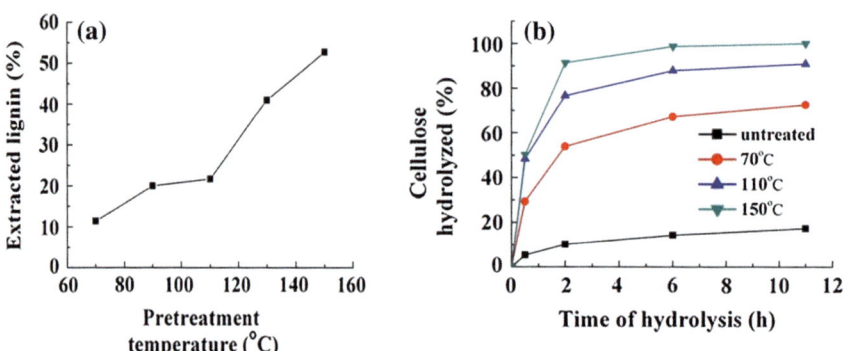

Fig. 6.7 Effect of pretreatment temperature on lignin extract rate from triticale straw by [C$_2$mim]OAc with a fixed pretreatment time of 1.5 h (**a**), and effect of pretreatment temperature on the hydrolysis of cellulose recovered from triticale straw after lignin removal by the ionic liquid(**b**) (Figure plotted with data from Fu et al. 2010)

Fig.6.8 Effect of pretreatment temperature on reducing sugar yield after enzyme hydrolysis using cellulase for wheat straw pretreated with [C$_2$mim]DEP 1 h (Figure plotted with data from Li et al. 2009)

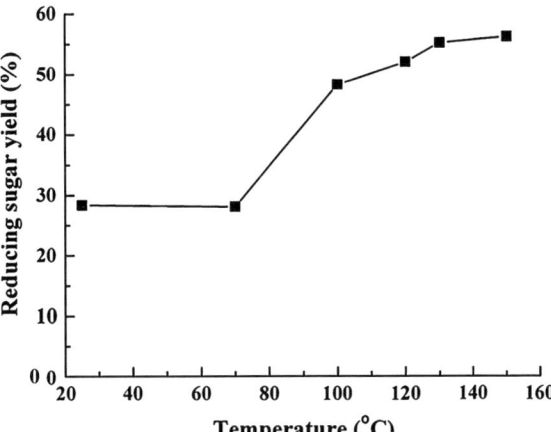

after enzyme hydrolysis of undissolved solid residue by a mixture of cellulase and β-glucosidase for 72 h at 47 °C. Figure 6.9 shows that 80 °C pretreatment temperature had almost no improvement over untreated sample, while 121 °C and 150 °C improved glucose and xylose yields considerably. For wheat straw, they found that 80 °C was ineffective and 121 °C showed only slight improvement, while 150 °C moretha ndouble dgluc osea ndxylos eyie lds.

Instead of conventional conductive heating to raise the temperature for enhanced biomass dissolution in ionic liquids, microwave irradiation can be used. Ha et al. (2011) found that microwave heating not only increased cotton cellulose solubility in ionic liquids, but also reduced the degree of polymerization in the regenerated cellulose obtained after the pretreatment compared with pretreatment at 110 °C without microwave. They suggested that internal heating by microwave irradiation was more effective for polar solvents such as ionic liquids. Casas et al. (2012) used microwave successfully as a thermal source to reduce dissolution time required for *Pinus radiata* and *Eucalyptus globulus* woods in the following ionic liquids: [C$_2$mim]OAc, [C$_4$mim]OAc, [C$_2$mim]Cl, [C$_4$mim]Cl, and [Amim]Cl. Despite its advantages, microwave heating also has significant drawbacks. Apart from increased equipment cost and scale-up difficulties, another drawback for microwave heating is that uneven heating may cause pyrolysis of cellulose due to high local temperatures (Feng and Chen 2008). Sonication was also used to enhance cellulose dissolution in ionic liquids by some researchers (Swatloski et al. 2002; El Seoud et al. 2007; Mikkola et al. 2007; Wang et al. 2012), but it is costly and difficult to scale up.

6.6 Pretreatment Time

Lignocellulosic biomass dissolution in an ionic liquid is far from instantaneous. Typically, several hours are required for biomass dissolution as shown in Table 6.4. Biomass solubility limit in a particular ionic liquid may not be reached within the

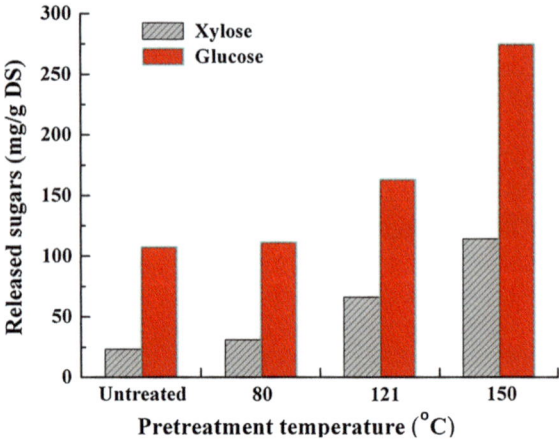

Fig.6.9 Effect of pretreatment temperature on glucose and xylose yields after [C$_2$mim]Clpre treated corn stover was enzyme hydrolyzed (Figure plotted with data from Pezoa et al. 2010)

given pretreatment time frame. The solubility of dissolving pulp cellulose in different ionic liquids was investigated by Swatloski et al. (2002). The highest solubility was obtained using [C$_4$mim]Cl. Various types of biomass were pretreated with [C$_2$mim]Cl at 150 °C by Pezoa et al. (2010) to study how lignin removal made the residual biomass easier to hydrolyze. Apart from lignin dissolution, some cellulose and hemicellulose were also dissolved. Thus, it was important to control pretreatment time. They found that for corn stover, wheat straw, and eucalyptus, 1 h pretreatment time led to significantly higher glucose and xylose yields compared with 0.5 h. However, the opposite trend was observed for Lenga (*Nothofagus pumilio*). They argued that the Lenga biomass pieces were smaller in size and thus faster for ionic liquid pretreatment. More cellulose and hemicellulose were lost due to dissolution in [C$_2$mim]Cl when the pretreatment time was excessive.

Xu et al. (2012) studied the effect of [C$_2$mim]OAc pretreatment time by dissolving corn stover at a relatively low temperature of 70 °C. Longer pretreatment time continuously improved sugar yield (Fig. 6.10). After 24 h of pretreatment, they achieved very good glucose and xylose yields of 84.9 % and 64.8 %, respectively. They observed that even after 24 h of pretreatment, undissolved residual corn stover was still found.

The effect of pretreatment time using triticale straw pretreatment in [C$_2$mim]OAc for lignin removal at 90 °C was investigated by Fu et al. (2010). Figure 6.11 shows that longer pretreatment time increased cellulose hydrolysis yield. A pretreatment time of 24 h was sufficient for complete cellulose hydrolysis. Obviously, pretreatment time and temperature are closely related. Higher temperature requires less time.

Comparing Figs. 6.7 and 6.11, at 150 °C pretreatment temperature only 1.5 h pretreatment time was required to achieve complete cellulose hydrolysis after pretreatment, while 90 °C required 24 h. In comparison with rice straw treated with [Cholinium]Lys at 90 °C for lignin removal, 5 h was considered the optimal pretreatment time as seen in Fig. 6.12 according to Hou et al. (2012). Lignin removal by [Cholinium]Lys greatly reduced the recalcitrance of rice straw. Without the pretreatment, the glucose and xylose yields from enzyme hydrolysis were only 20.4 and6.8%,re spectively.

When a temperature of 130 °C was used in the pretreatment of wheat straw in [C$_2$mim]DEP, pretreatment times greater than 30 min showed very little improvements as shown in Fig. 6.13 (Li et al. 2009). This was because 130 °C was a sufficientlyhighpre treatmentte mperaturea sindic atede arlieri nFi g.6.8.

6.7 WaterC ontent

Ionic liquids tend to be highly hygroscopic (Brandt et al. 2012), which can bring water into the pretreatment process. Water can also be introduced by moist biomass. Mazza et al. (2009) investigated the effect of water on cellulose dissolution

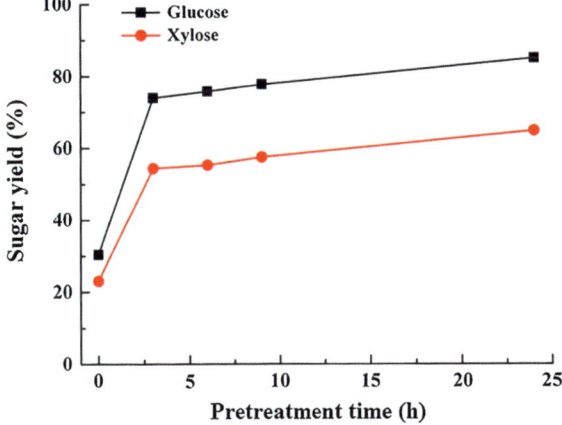

Fig. 6.10 Effect of pretreatment time on sugar yields after enzyme hydrolysis using cellulase for corn stover after pretreatment with [C$_2$mim]OAc at 70 °C (Figure plotted with data from Xu et al. 2012)

in [C$_4$mim]Cl and 1,3-dimethylimidazolium dimethylphosphate ([C$_1$mim]DMP).

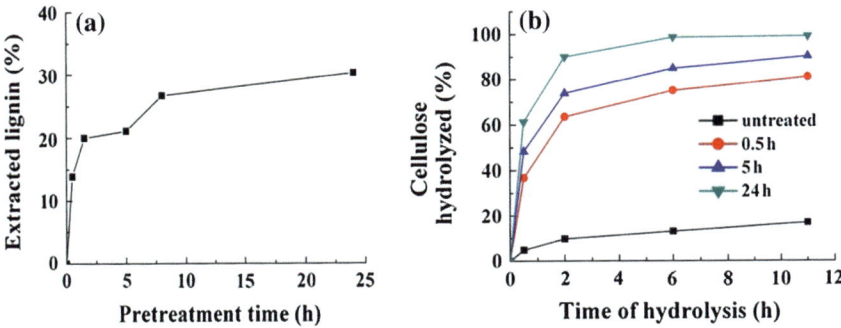

Fig. 6.11 Effect of pretreatment time on lignin extract rate in pretreatment of triticale straw with [C$_2$mim]OAc at 90 °C (**a**), and effect of pretreatment time on enzyme hydrolysis of the cellulose recovered from triticale straw after lignin removal by the ionic liquid (**b**) (Figure plotted with data from Fu et al. 2010)

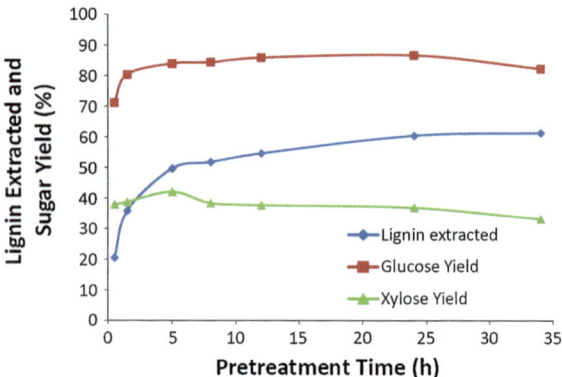

Fig.6.12 Effect of pretreatment time on lignin extraction and sugar yield after enzyme hydrolysis using cellulase for rice straw pretreated with [Cholinium]Lysa t90 °C (Figure plotted with data fromH oue ta l.2012)

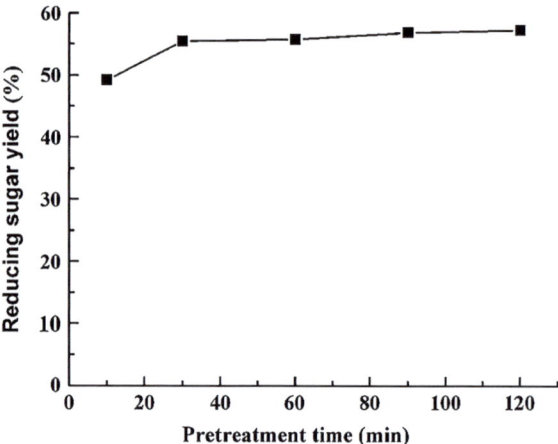

Fig.6.13 Effect of pretreatment time on reducing sugar yield after enzyme hydrolysis of wheat straw pretreated with [C$_2$mim]DEP at 130 °C (Figure plotted with data fromLi e ta l.2009)

They found that increasing water content progressively reduced cellulose solubility. Adding a small amount of water to a cellulose solution in [C$_1$mim]DMP formed cellulose aggregates. This is consistent with the finding by Swatloski et al. (2002) that cellulose solubility decreased significantly in some ionic liquids when a small amount of water was present. They argued that this probably was a result of competitive hydrogen bonding. Interestingly, Brandt et al. (2010) observed that a small amount of water promoted lignin (in pine) solubilization in [C$_4$mim]Cl, but not in [C$_4$mim]OAc. It was likely that the presence of water resulted in the release of HCl acid that was beneficial according to Xie and Shi (2006).

Water can be removed from ionic liquids using vacuum at an elevated temperature (Tan et al. 2009). One traditional method is to dry an ionic liquid over a mixture of Na$_2$SO$_4$ and MgSO$_4$ before vacuum drying (Kilpeläinen et al. 2007; Shi et al. 2008). Freeze drying has also been used (Vitz et al. 2009). Furthermore, water can be stripped from an ionic liquid by sparging dry nitrogen gas through the ionic liquid to a mass fraction as low as 0.001 within a few hours (Ren et al. 2010). Biomass drying has been well documented in the literature.

6.8 IonicL iquidR ecyclingandB iomassR ecovery

Due to the high cost and the potential toxicity of ionic liquids, they must be recycled after biomass pretreatment. Various methods have been used by researchers to recover ionic liquids after biomass pretreatment. Each method has its advantages and limitations (Tan and MacFarlane 2010).

6.8.1 PrecipitationU sing Antisolvents

Cellulose can be recovered from its ionic liquid solution by adding an antisolvent such as water, which precipitates cellulose from the ionic liquid (Zavrel et al. 2009). Subsequently, the precipitated cellulose can be regenerated using filtration or centrifugation. However, cellulose precipitates typically appear as a gel, making separation difficult. Sometimes, a water/acetone (1:1 v/v) mixture is used instead of water to avoid gel formation (Shill et al. 2011).

6.8.2 DistillationandE vaporation

Distillation or evaporation can be used to remove volatile antisolvents easily because most ionic liquids are practically nonvolatile and they have good thermo-stabilities. However, this is usually an energy-intensive process. Direct removal of ionic liquid without the antisolvent step is also possible. Earle et al. (2006) demonstrated that some ionic liquids are distillable at low pressure without decomposition, but relatively high temperatures (200–300 °C) are needed. Olivier-Bourbigou et al. (2010) described several routes to convert ionic liquids into easily distillable compounds that can be subsequently converted back to ionic liquids by chemical reactions. This obviously could complicate process design.

6.8.3 PhaseSe parationandLiquid–LiquidP artitioning

Another method for ionic liquid recovery from the pretreatment process is to utilize their ability to form a biphasic system by adding an aqueous solution containing phosphate, carbonate, or sulfate anion, etc. For example, adding potassium phosphate to a water and [C$_4$mim]Cl mixture forms an aqueous biphasic system from which the ionic liquid can be recovered (Shill et al. 2011). Deng et al. (2009) found that the recovery of [Amim]Cl is better using a salt with a higher salting-out strength (e.g., K$_3$PO$_4$ > K$_2$HPO$_4$ > K$_2$CO$_3$). Using 46.5 % K$_2$HPO$_4$ concentration, they achieved [Amim]Cl recovery as high as 96.8 %. Phase equilibrium data for various ionic liquid/ water/salt systems have been presented by several research groups (Deng et al. 2009; Pei et al. 2009; Shill et al. 2011). Alcohols and supercritical CO$_2$ may also be used for phase partitioning of ionic liquid systems (Aki et al. 2004; Crosthwaite et al. 2004).

6.8.4 MembraneF iltration

Precipitates from ionic liquid pretreatment of biomass can be separated using membrane filtration. For example, nanofiltration membranes have been used to reject ionic liquids while allowing smaller molecules to pass through for ionic liquid recycling (Han et al. 2005; Hazarika et al. 2012). Membrane filtration has the advantage that room temperature operation may be possible. However, high viscosity is often a hurdle.

6.8.5 Chromatography

Ion exclusion chromatography, which is already in use for carbohydrate processing, can also be used for ionic liquid recovery because it can exclude charged species such as ionic liquid species, thus separating them from nonelectrolytes such as sugar molecules (Binder and Raines 2010). For continuous chromatographic separation, Simulated Moving Bed (SMB) chromatography is used. SMB utilizes a series of ion-exchange chromatography columns with synchronized moving ports to simulate resin movement. Although chromatography separation has the advantage of high resolution and more than two separated fractions in the output, it is far more expensive than other separation methods.

6.9 MobileandOn-far mPr etreatmentU singI onicL iquids

Most lignocellulosic biomass tend to be bulky and not cost-effective for transportation from remote agricultural areas to a central processing facility. Mobile and on-farm biomass processing may be an answer to address this problem. Lignocellulosic biomass may be processed using a truck-mounted or on-farm-scale processor to obtain intermediary products such as fermentable sugars that can be shipped to a centralized facility for further processing or fermentation. A self-contained complete mobile or on-farm processor may even be used to produce bioethanol as a fuel cost-effectively in some regions that have abnormally high fuel costs such as a remote military forward operating base in a war zone. Mobile and on-farm processing units favor green pretreatment methods to eliminate the need for on-site waste treatment.

Unlike conventional organic solvents that are volatile, ionic liquids do not give out measurable vapor pressure. This makes them "green" solvents because they do not pollute the atmosphere (Blanchard et al. 1999). Ionic liquids can be recycled easily after pretreatment, thus no waste treatment is needed after the pretreatment process. A pretreatment process that uses ionic liquid typically uses mild temperature and atmosphere pressure without harsh conditions. Thus, ionic liquid pretreatment is particularly suitable for mobile and on-farm lignocellulosic biomass processing.

Figure 6.14 shows a flowchart used by Shill et al. (2011) to process gram quantities of Avicel and *Miscanthus giganteus* (a large perennial grass). The highest temperature used in the process was 140 °C (for biomass dissolution) that is much lower than the 250 °C used in steam pretreatment by Kim and Hong (2001). Figure 6.15 shows a process designed by Sen et al. (2012) for ionic liquid pretreatment of corn stover with ionic liquid recycling using an SMB chromatographic system instead of evaporation of the antisolvent (such as water in the previous flowchart) because evaporation is an energy-intensive process. However, the SMB system overtook the cost of the ionic liquid as the most expensive cost factor in capital investment for the large-scale operation (Sen et al. 2012). SMB may be replaced by a liquid–liquid partitioning process to reduce cost. The process shown in Fig. 6.16 used by Sun et al. (2009) may be scaled up for mobile or on-farm application. It uses water and acetone to selectively precipitate cellulose and lignin from softwood and hardwood after the wood is completely dissolved in [C_2mim]OAc. The ionic liquid is recycled in the process. A less volatile solvent may be used to replace acetone for lignin recovery if desired. All these pretreatment processes may be refined and scaled up or scaled down properly for mobile andon-f arma pplications.

6.10 Conclusion

In recent years, ionic liquid pretreatment of lignocellulosic biomass has drawn considerable attention. They are effective in cellulose dissolution for many types of lignocellulosic biomass. Some ionic liquids can also be used to selectively dissolve lignin and hemicellulose in addition to cellulose, and the components can be selectively recovered for hydrolysis. Separation of cellulose from lignin and hemicellulose greatly enhances cellulose hydrolysis after the pretreatment because the extensive networking consisting of lignin, hemicellulose, and cellulose can be disrupted by the ionic liquid pretreatment. Recovery of dissolved biomass components is relatively easy and ionic liquids can be recycled. Due to their solvation ability and basicity or acidity, ionic liquids are also special solvents that can be used for bioconversion to produce renewable bioproducts as specialty chemicals.

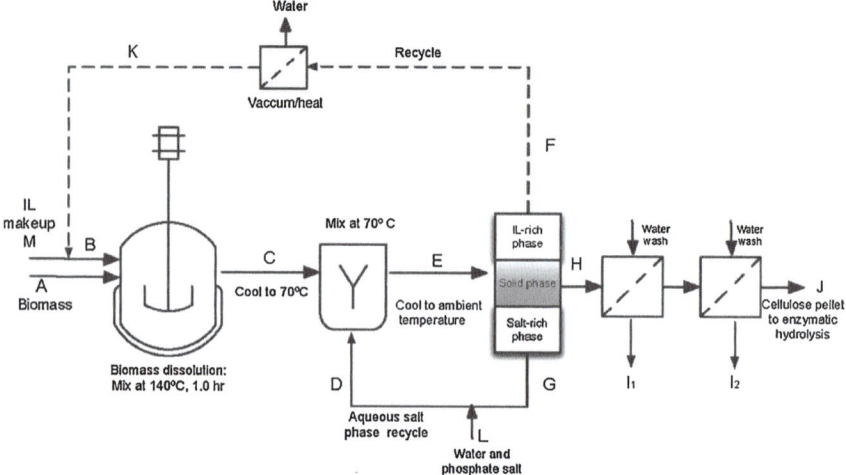

Fig. 6.14 A lignocellulosic biomass pretreatment process using an ionic liquid for cellulose dissolution. (Reprinted from Biotechnology and Bioengineering, Ionic liquid pretreatment of cellulosic biomass: Enzymatic hydrolysis and ionic liquid recycle, Kierston Shill, Sasisanker Padmanabhan, Qin Xin, John M.Prausnitz, Douglas S. Clark, Harvey W. Blanch, 108:511–520, Copyright 2010, with permission from John Wiley and Sons)

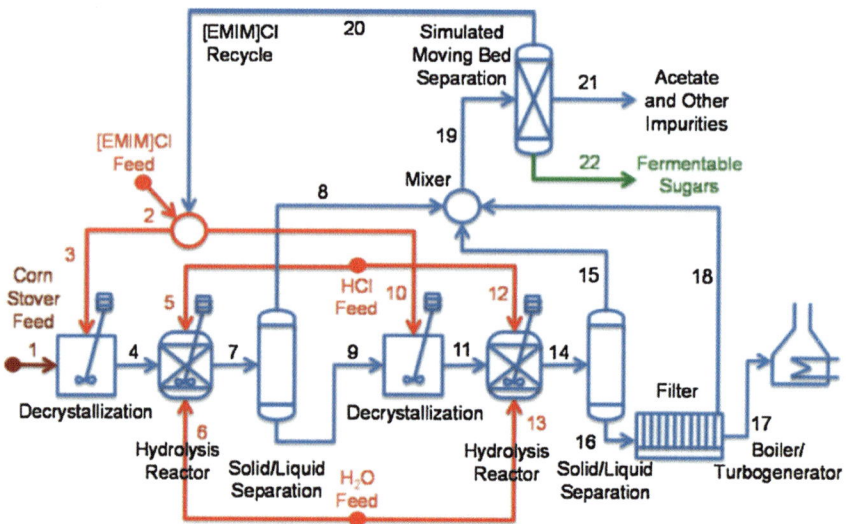

Fig. 6.15 A proposed corn stover pretreatment process using [C2mim]Cl that incorporates SMB for ionic liquid recycling. (Reprinted from Biofuels, Bioproducts and Biorefining, Conversion of biomass to sugars via ionic liquid hydrolysis: process synthesis and economic evaluation, S. Murat Sen, Joseph B. Binder, Ronald T. Raines, Christos T. Maravelias, 6:444–452, Copyright 2010, with permission from John Wiley and Sons)

Fig.6.16 A process with ionic liquid recycle for cellulose and lignin recovery after complete dissolution of softwood or hardwood in [C₂mim]OAc(A daptedf rom Sun et al. 2009)

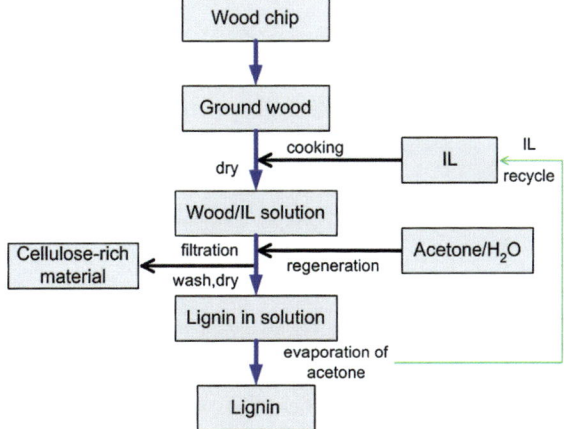

Although attractive, ionic liquid pretreatment of lignocellulosic biomass is still not deployed for practical applications because of several drawbacks. No pilot-scale tests have been reported so far. The most noticeable obstacle is the initial high chemical cost. Although research is underway using amino acids to make biodegradable ionic liquids, most ionic liquids for biomass processing are not biodegradable. Toxicity, corrosivity, and hygroscopicity of ionic liquids are also major concerns. Ionic liquids can be recycled easily, but a tiny amount unrecovered ionic liquids may still present a safety concern if the recycling is not complete. Residual ionic liquids in the biomass may cause enzyme inhibition in the hydrolysis step and can be toxic to the microorganisms in the fermentation step (Alvira et al. 2010). Despite these hurdles, ionic liquids have a promising future for biomass pretreatment and processing. It can be envisioned that in the not-too-distant future some low-cost biodegradable ionic liquids with no or low toxicity will be available at reasonable costs. This would make their practical deployment feasible for mobile, on-farm, or even large-scale applications.

References

Aki SNVK, Mellein BR, Saurer EM, Brennecke JF (2004) High-pressure phase behavior of carbon dioxide with imidazolium-based ionic liquids. J Phys Chem B 108:20355–20365

Alizadeh H, Teymouri F, Gilbert TI, Dale BE (2005) Pretreatment of switchgrass by ammonia fibere xplosion(AFEX). ApplB iochemB iotechnol124: 1133–1141

Alvira P, Tomás-Pejó E, Ballesteros M, Negro MJ (2010) Pretreatment technologies for an efficient bioethanol production process based on enzymatic hydrolysis: A review. Bioresour Technol101: 4851–4861

Baker GA, Baker SN, Pandey S, Bright FV (2005) An analytical view of ionic liquids. Analyst 130:800–808

Binder JB, Raines RT (2010) Fermentable sugars by chemical hydrolysis of biomass. Proc Natl AcadSc i107:451 6–4521

Blanchard LA, Brennecke JF (2001) Recovery of organic products from ionic liquids using supercritical carbon dioxide. IndE ngC hemR es40: 287–292

Blanchard LA, Hancu D, Beckman EJ, Brennecke JF (1999) Green processing using ionic liquids and CO_2. Nature 399:28–29

Brandt A, Hallett JP, Leak DJ, Murphy RJ, Welton T (2010) The effect of the ionic liquid anion in the pretreatment of pine wood chips. Green Chem 12:672

Brandt A, Erickson JK, Hallett JP, Murphy RJ, Potthast A, Ray MJ, Rosenau T, Schrems M, Welton T (2012) Soaking of pine wood chips with ionic liquids for reduced energy input during grinding. Green Chem14: 1079–1085

Casas A, Oliet M, Alonso MV, Rodríguez F (2012) Dissolution of *Pinus radiata* and *Eucalyptus globulus* woods in ionic liquids under microwave radiation: lignin regeneration and characterization. Sep Purif Technol9 7:115–122 http://dx.doi.org/10.1016/j.seppur.2011.12.032

Çetinkol ÖP, Dibble DC, Cheng G, Kent MS, Knierim B, Auer M, Wemmer DE, Pelton JG, Melnichenko YB, Ralph J et al (2010) Understanding the impact of ionic liquid pretreatment on eucalyptus. Biofuels1: 33–46

Cheng G, Varanasi P, Li C, Liu H, Melnichenko YB, Simmons BA, Kent MS, Singh S (2011) Transition of cellulose crystalline structure and surface morphology of biomass as a function of ionic liquid pretreatment and its relation to enzymatic hydrolysis. Biomacromolecules 12:933–941

Crosthwaite JM, Aki SNVK, Maginn EJ, Brennecke JF (2004) Liquid phase behavior of imidazolium-based ionic liquids with alcohols. J PhysC hemB 1 08:5113–5119

Cull SG, Holbrey JD, Vargas-Mora V, Seddon KR, Lye GJ (2000) Room-temperature ionic liquids as replacements for organic solvents in multiphase bioprocess operations. Biotechnol Bioeng69:227–233

Deng Y, Long T, Zhang D, Chen J, Gan S (2009) Phase diagram of [Amim] Cl + salt aqueous biphasic systems and its application for [Amim]Cl recovery†. J Chem Eng Data 54:2470–2473

Earle MJ, Esperança JMSS, Gilea MA, Lopes JNC, Rebelo LPN, Magee JW, Seddon KR, Widegren JA (2006) The distillation and volatility of ionic liquids. Nature 439:831–834

El Seoud OA, Koschella A, Fidale LC, Dorn S, Heinze T (2007) Applications of ionic liquids in carbohydrate chemistry: a window of opportunities. Biomacromolecules 8:2629–2647

Erdmenger T, Haensch C, Hoogenboom R, Schubert US (2007) Homogeneous tritylation of cellulose in 1-Butyl-3-methylimidazolium chloride. MacromolB iosci7: 440–445

Feng L, Chen Z (2008) Research progress on dissolution and functional modification of cellulose in ionic liquids. J Mol Liq142: 1–5

Fort DA, Remsing RC, Swatloski RP, Moyna P, Moyna G, Rogers RD (2007) Can ionic liquids dissolve wood? Processing and analysis of lignocellulosic materials with 1-n-butyl-3-methylimidazolium chloride. Green Chem9: 63–69

Fu D, Mazza G (2011) Optimization of processing conditions for the pretreatment of wheat straw using aqueous ionic liquid. Bioresour Technol102: 8003–8010

Fu D, Mazza G, Tamaki Y (2010) Lignin extraction from straw by ionic liquids and enzymatic hydrolysis of the cellulosic residues. J Agric Food Chem 58:2915–2922

Fukaya Y, Hayashi K, Wada M, Ohno H (2008) Cellulose dissolution with polar ionic liquids under mild conditions: required factors for anions. Green Chem 10:44–46

Garrote G, Dominguez H, Parajo JC (1999) Hydrothermal processing of lignocellulosic materials. Eur J Wood Wood Prod 57:191–202

Graenacher C (1934) Cellulose solution. U.S. Patent 1,943,176

Ha SH, Mai NL, An G, Koo YM (2011) Microwave-assisted pretreatment of cellulose in ionic liquid for accelerated enzymatic hydrolysis. Bioresour Technol 102:1214–1219

HanX , Armstrong DW(2007)I onicI iquidsi ns eparations. AccC hemR es40: 1079–1086

Han S, Wong HT, Livingston AG (2005) Application of organic solvent nanofiltration to separation of ionic liquids and products from ionic liquid mediated reactions. Chem Eng Res Des 83:309–316

Hazarika S, Dutta NN, Rao PG (2012) Dissolution of lignocellulose in ionic liquids and its recovery by nanofiltration membrane. Sep Purif Technol 97:123–129 http://dx.doi.org /10.1016/j.seppur.2012.04.026

Hendriks A, Zeeman G (2009) Pretreatments to enhance the digestibility of lignocellulosic bio-mass.B ioresour Technol100: 10–18

Hou XD, Smith TJ, Li N, Zong MH (2012) Novel renewable ionic liquids as highly effective solvents for pretreatment of rice straw biomass by selective removal of lignin. Biotechnol Bioeng109:2484–2493doi :10.1002/ bit.24522(inpr ess)

Huddleston JG, Rogers RD (1998) Room temperature ionic liquids as novel media for "clean"liquid–liquid extraction. ChemC ommun16: 1765–1766

Kaar W, Gutierrez C, Kinoshita C (1998) Steam explosion of sugarcane bagasse as a pretreat-ment for conversion to ethanol. Biomass Bioenergy 14:277–287

Karunanithy C, Muthukumarappan K (2011) Optimization of switchgrass and extruder parameters for enzymatic hydrolysis using response surface methodology. Ind Crops Prod 33:188–199

Kilpeläinen I, Xie H, King A, Granstrom M, Heikkinen S, Argyropoulos DS (2007) Dissolution of wood in ionic liquids. J Agric Food Chem 55:9142–9148

Kim KH, Hong J (2001) Supercritical CO_2 pretreatment of lignocellulose enhances enzymatic cellulose hydrolysis. Bioresour Technol77: 139–144

Kim TH, Lee Y (2007) Pretreatment of corn stover by soaking in aqueous ammonia at moderate temperatures. ApplB iochemB iotechnol137: 81–92

Klemm D, Heublein B, Fink H-P, Bohn A (2005) Cellulose: fascinating biopolymer and sustain-able raw material. Angew Chem Int Ed 44:3358–3393

Lan W, Liu CF, Sun RC (2011) Fractionation of bagasse into cellulose, hemicelluloses and lignin with ionic liquid treatment followed by alkaline extraction. J Agric Food Chem 59:8691–8701

Laureano-Perez L, Teymouri F, Alizadeh H, Dale BE (2005) Understanding factors that limit enzymatic hydrolysis of biomass. Springer, Netherlands, pp 1081–1099

Laus G, Bentivoglio G, Schottenberger H, Kahlenberg V, Kopacka H, Röder T, Sixta H (2005) Ionic liquids: current developments, potential and drawbacks for industrial applications. LenzingerB erichte84: 71–85

Lee SH, Doherty TV, Linhardt RJ, Dordick JS (2009) Ionic liquid-mediated selective extraction of lignin from wood leading to enhanced enzymatic cellulose hydrolysis. Biotechnol Bioeng 102:1368–1376

Li C, Wang Q, Zhao ZK (2008) Acid in ionic liquid: an efficient system for hydrolysis of ligno-cellulose. Green Chem10: 177–182

Li Q, He YC, Xian M, Jun G, Xu X, Yang JM, Li LZ (2009) Improving enzymatic hydrolysis of wheat straw using ionic liquid 1-ethyl-3-methyl imidazolium diethyl phosphate pretreatment. Bioresour Technol100: 3570–3575

Li C, Knierim B, Manisseri C, Arora R, Scheller HV, Auer M, Vogel KP, Simmons BA, Singh S (2010) Comparison of dilute acid and ionic liquid pretreatment of switchgrass: bio-mass recalcitrance, delignification and enzymatic saccharification. Bioresour Technol 101:4900–4906

Li C, Cheng G, Balan V, Kent MS, Ong M, Chundawat SPS, daCosta Sousa L, Melnichenko YB, Dale BE, Simmons BA, Singh S (2011a) Influence of physico-chemical changes on enzymatic digestibility of ionic liquid and AFEX pretreated corn stover. Bioresour Technol 102:6928–6936

Li W, Sun N, Stoner B, Jiang X, Lu X, Rogers RD (2011b) Rapid dissolution of lignocellulosic biomass in ionic liquids using temperatures above the glass transition of lignin. Green Chem 13:2038–2047

Liu H, Sale KL, Holmes BM, Simmons BA, Singh S (2010) Understanding the interactions of cellulose with ionic liquids: a molecular dynamics study. J Phys Chem B 114:4293–4301

Liu QP, Hou XD, Li N, Zong MH (2012) Ionic liquids from renewable biomaterials: synthesis, characterization and application in the pretreatment of biomass. Green Chem14: 304–307

Lloyd TA, Wyman CE (2005) Combined sugar yields for dilute sulfuric acid pretreatment of corn stover followed by enzymatic hydrolysis of the remaining solids. Bioresour Technol 96:1967–1977

Lynam JG, Toufiq Reza M, Vasquez VR, Coronella CJ (2012) Pretreatment of rice hulls by ionic liquid dissolution. Bioresour Technol114: 629–636

Mäki-Arvela P, Anugwom I, Virtanen P, Sjöholm R, Mikkola JP (2010) Dissolution of lignocellulosic materials and its constituents using ionic liquids—a review. Ind Crops Prod 32:175–201

Mamman AS, Lee JM, Kim YC, Hwang IT, Park NJ, Hwang YK, Chang JS, Hwang JS (2008) Furfural: hemicellulose/xylosederivedbi ochemical.B iofuelsB ioprodB iorefin2: 438–454

Marsh KN, Boxall JA, Lichtenthaler R (2004) Room temperature ionic liquids and their mixtures—a review. Fluid Phase Equilib 219:93–98

Mazza M, Catana D-A, Vaca-Garcia C, Cecutti C (2009) Influence of water on the dissolution of cellulose in selected ionic liquids. Cellulose 16:207–215

Mikkola JP, Kirilin A, Tuuf JC, Pranovich A, Holmbom B, Kustov LM, Murzin DY, Salmi T (2007) Ultrasound enhancement of cellulose processing in ionic liquids: from dissolution towards functionalization. Green Chem 9:1229–1237

Mosier N, Wyman C, Dale B, Elander R, Lee Y, Holtzapple M, Ladisch M (2005) Features of promising technologies for pretreatment of lignocellulosic biomass. Bioresour Technol 96:673–686

Narayanaswamy N (2010) Supercritical carbon dioxide pretreatment of various lignocellulosic biomasses. MS Thesis, Ohio University, Athens

Narayanaswamy N, Faik A, Goetz DJ, Gu T (2011) Supercritical carbon dioxide pretreatment of corn stover and switchgrass for lignocellulosic ethanol production. Bioresour Technol 102:6995–7000

Novoselov NP, Sashina ES, Petrenko VE, Zaborsky M (2007) Study of dissolution of cellulose in ionic liquids by computer modeling. Fibre Chem39: 153–158

Olivier-Bourbigou H, Magna L, Morvan D (2010) Ionic liquids and catalysis: Recent progress fromkno wledgetoa pplications. ApplC atal A373: 1–56

Pei Y, Wang J, Wu K, Xuan X, Lu X (2009) Ionic liquid-based aqueous two-phase extraction of selected proteins. Sep Purif Technol 64:288–295

Pezoa R, Cortinez V, Hyvarinen S, Reunanen J, Linenqueo ME, Salazar O, Carmona R, Garcia A, Murzin DY, Mikkola J-P (2010) Use of ionic liquids in the pretreatment of forest and agricultural residues for the production of bioethanol.C ellulC hem Technol44: 165–172

Plechkova NV, Seddon KR (2008) Applications of ionic liquids in the chemical industry. Chem Soc Rev 37:123–150

Remsing RC, Swatloski RP, Rogers RD, Moyna G (2006) Mechanism of cellulose dissolution in the ionic liquid 1-n-butyl-3-methylimidazolium chloride: a 13C and 35/37Cl NMR relaxation study on model systems. ChemC ommun12: 1271–1273

Ren S, Hou Y, Wu W, Liu W (2010) Purification of ionic liquids: sweeping solvents by nitrogen. JC hemEng D ata55: 5074–5077

Runge CF, Senauer B (2007) How biofuels could starve the poor. Foreign Aff 86 (May/June issue)41

Schultz TP, Biermann CJ, McGinnis GD (1983) Steam explosion of mixed hardwood chips as a biomass pretreatment. IndE ngC hemP rodR esD ev22: 344–348

Sen SM, Binder JB, Raines RT, Maravelias CT (2012) Conversion of biomass to sugars via ionic liquid hydrolysis: process synthesis and economic evaluation. Biofuels Bioprod Biorefin 6:444–452

Shi YG, Fang Y, Ren YP, Wu HP, Guan HL (2008) Effect of ionic liquid [BMIM][PF$_6$] on asymmetric reduction of ethyl 2-oxo-4-phenylbutyrate by Saccharomyces cerevisiae. J Ind Microbiol Biotechnol 35:1419–1424

Shill K, Padmanabhan S, Xin Q, Prausnitz JM, Clark DS, Blanch HW (2011) Ionic liquid pretreatment of cellulosic biomass: enzymatic hydrolysis and ionic liquid recycle. Biotechnol Bioeng108:511–5 20

Singh S, Simmons BA, Vogel KP (2009) Visualization of biomass solubilization and cellulose regeneration during ionic liquid pretreatment of switchgrass. Biotechnol Bioeng 104:68–75

Sun N, Rahman M, Qin Y, Maxim ML, Rodríguez H, Rogers RD (2009) Complete dissolution and partial delignification of wood in the ionic liquid 1-ethyl-3-methylimidazolium acetate. Green Chem11:646–655

Sun N, Rodríguez H, Rahman M, Rogers RD (2011) Where are ionic liquid strategies most suited in the pursuit of chemicals and energy from lignocellulosic biomass? Chem Commun 47:1405–1421

Swatloski RP, Spear SK, Holbrey JD, Rogers RD (2002) Dissolution of cellose with ionic liquids. J Am Chem Soc 124:4974–4975

Tan S, MacFarlane D (2010) Ionic liquids in biomass processing. Ion Liq 290:311–339

Tan SSY, MacFarlane DR, Upfal J, Edye LA, Doherty WOS, Patti AF, Pringle JM, Scott JL (2009) Extraction of lignin from lignocellulose at atmospheric pressure using alkylbenzene-sulfonate ionic liquid. Green Chem11: 339–345

Teymouri F, Laureano-Perez L, Alizadeh H, Dale BE (2004) Ammonia fiber explosion treatment of corn stover. Appl Biochem Biotechnol 115:951–963

Vitz J, Erdmenger T, Haensch C, Schubert US (2009) Extended dissolution studies of cellulose in imidazolium based ionic liquids. Green Chem 11:417–424

Vo HT, Kim CS, Ahn BS, Kim HS, Lee H (2011) Study on dissolution and regeneration of poplar woodin imidazolium-basedi onicl iquids.J WoodC hem Technol31: 89–102

Wan C, Li Y (2010) Microbial pretreatment of corn stover with *Ceriporiopsis subvermispora* for enzymatic hydrolysis and ethanol production. Bioresour Technol101: 6398–6403

Wang H, Gurau G, Rogers RD (2012) Ionic liquid processing of cellulose. Chem Soc Rev 41:1519–1537

Weerachanchai P, Leong SSJ, Chang MW, Ching CB, Lee JM (2012) Improvement of biomass properties by pretreatment with ionic liquids for bioconversion process. Bioresour Technol 111:453–459

XieH ,Shi T(2006) Woodl iquefactionbyi onicl iquids.H olzforschung60: 509–512

Xin Q, Pfeiffer K, Prausnitz JM, Clark DS, Blanch HW (2012) Extraction of lignins from aqueous–ionic liquid mixtures by organic solvents. Biotechnol Bioeng 109:346–352

Xu A, Wang J, Wang H (2010) Effects of anionic structure and lithium salts addition on the dissolution of cellulose in 1-butyl-3-methylimidazolium-based ionic liquid solvent systems. Green Chem12:268–275

Xu F, Shi YC, Wang D (2012) Enhanced production of glucose and xylose with partial dissolution of corn stover in ionic liquid, 1-Ethyl-3-methylimidazolium acetate. Bioresour Technol 114:720–724

Yang B, Wyman CE (2008) Pretreatment: the key to unlocking low-cost cellulosic ethanol. BiofuelsB ioprod Biorefin2: 26–40

Yu G, Yano S, Inoue H, Inoue S, Endo T, Sawayama S (2010) Pretreatment of rice straw by a hot-compressed water process for enzymatic hydrolysis. Appl Biochem Biotechnol 160:539–551

Zavrel M, Bross D, Funke M, Büchs J, Spiess AC (2009) High-throughput screening for ionic liquids dissolving (ligno-) cellulose. Bioresour Technol100: 2580–2587

Zhang J, Zhang H, Wu J, Zhang J, He J, Xiang J (2010) NMR spectroscopic studies of cellobiose solvation in EmimAc aimed to understand the dissolution mechanism of cellulose in ionic liquids.Phys C hemC hemPhys 1 2:1941–1947

Zhao H, Baker GA, Song Z, Olubajo O, Crittle T, Peters D (2008) Designing enzyme-compatible ionic liquids that can dissolve carbohydrates. Green Chem10: 696–705

Zhao H, Jones CL, Baker GA, Xia S, Olubajo O, Person VN (2009) Regenerating cellulose from ionic liquids for an accelerated enzymatic hydrolysis. J Biotechnol 139:47–54

Zheng Y, Lin HM, Wen J, Cao N, Yu X, Tsao GT (1995) Supercritical carbon dioxide explosion as a pretreatment for cellulose hydrolysis. Biotechnol Lett 17:845–850

Zhu S, Wu Y, Chen Q, Yu Z, Wang C, Jin S, Ding Y, Wu G (2006) Dissolution of cellulose with ionic liquids and its application: a mini-review. Green Chem 8:325–327

Zhu Z, Zhu M, Wu Z (2012) Pretreatment of sugarcane bagasse with NH_4OH-H_2O_2 and ionic liquid for efficienth ydrolysisa ndbi oethanolp roduction.B ioresour Technol1 19:199–207

2231995R00097

Printed in Germany
by Amazon Distribution
GmbH, Leipzig